Preface

NooJ is a linguistic development environment that provides tools for linguists to construct linguistic resources that formalize a large gamut of linguistic phenomena: typography, orthography, lexicons for simple words, multiword units and discontinuous expressions, inflectional and derivational morphology, local, structural and transformational syntax, and semantics.

For each resource that linguists create, NooJ provides parsers that can apply it to any corpus of texts and extract examples or counter-examples, annotate matching sequences, perform statistical analyses, and so on. NooJ also contains generators that can produce the texts that these linguistic resources describe, as well as serving as a rich toolbox that allows linguists to construct, maintain, test, debug, accumulate, and reuse linguistic resources.

For each elementary linguistic phenomenon to be described, NooJ proposes a set of computational formalisms, the power of which ranges from very efficient finite-state automata to very powerful Turing machines. This makes NooJ's approach different from most other computational linguistic tools that typically offer a unique formalism to their users.

Since its first release in 2002, NooJ has been enhanced with new features every year. Linguists, researchers in social sciences, and more generally all professionals who analyze texts have contributed to its development and participated in the annual NooJ conference. In 2013, a new version of NooJ was released, based on the JAVA technology and available to all as an open source GPL project. Moreover, several private companies are now using NooJ to construct business applications in several domains, from business intelligence to opinion analysis.

At the opening of the International NooJ 2016 Conference, which was held during June 9–11 at the University of South Bohemia in České Budějovice (Czech Republic), Max Silberztein presented his book *Formalizing Natural Languages: The NooJ Approach* (Wiley, 2016), which describes the theoretical and methodological backgrounds of the NooJ approach. The present volume contains 21 articles selected from the other 32 papers presented at the conference. These articles are organized in four parts: "Vocabulary and Morphology" containing seven articles; "Corpus Processing and Information Extraction" containing six articles; "Syntactic Analysis" containing five articles; and "Semantic Analysis and Its Applications" containing three articles.

The articles in the first part cover the construction of electronic dictionaries, the description of phonetic and morphological information, and the construction of bilingual electronic dictionaries that can be used by machine translation software.

– Hamid Annouz's article "Treatments of the Kabylian Derived Nominal Verbs with NooJ" describes the formalization of nominalization of Kabylian verbs in the context of a full-coverage description of the Kabylian vocabulary.

- Stanislau Lysy, Hanna Stanislavenka, and Yury Hetsevich's article "Addition of IPA Transcription to the Belarusian NooJ Module" shows how to integrate phonetic information (including stress position) in a NooJ dictionary.
- Kristina Kocijan, Marijana Janjić, and Sara Librenjak's article "Recognizing Diminutive and Augmentative Croatian Nouns" presents a set of morphological grammars used to recognize Croatian diminutive and augmentative nouns, using the base nouns listed in the Croatian dictionary.
- Dhekra Najar, Slim Mesfar, and Henda Ben Ghezela's article "Inflectional and Morphological Variation of Arabic Multi-Word Expressions" presents a linguistic module that includes dictionaries, morphological and syntactic grammars, used to recognize multi-word units in Arabic texts.
- Maximilian Duran's article "Quechua Module for NooJ Multilingual Linguistic Resources for MT" presents the first available electronic dictionary for Quechua that formalizes inflectional and derivational morphology.
- Francesca Esposito and Annibale Elia's article "NooJ Local Grammars for Innovative Start-Up Language" presents a system capable of analyzing the vocabulary of specialized texts such as the documents of start-up companies.
- Hager Cheikhrouhou's article "Arabic Translation of the French Auxiliary: Using the Platform NooJ" uses the class "X" of auxiliary verbs from the Dubois LVF dictionary and presents the corresponding dictionary for Arabic verbs.

The articles in the second part involve the construction of automatic software capable of parsing and extracting meaningful information from texts by retrieving terms or named entities automatically:

- Nadia Ghezaiel Hammouda and Kais Haddar's article "Integration of a Segmentation Tool for Arabic Corpora in NooJ Platform to Build an Automatic Annotation Tool" presents a set of grammars used to recognize and annotate sentences from Arabic texts.
- Yury Hetsevich, V. Varanovich, E. Kachan, I. Reentovich, and S. Lysy's article "Semi-automatic Part-of-Speech Annotating for Belarusian Dictionaries Enrichment in NooJ" presents an algorithm that uses a corpus of one million words to automatically enrich the NooJ Belarusian dictionary.
- Francesca Parisi's article "Clinical Term Recognition: From Local to LOINC Terminology. An Application for Italian Language" uses a clinical and biological terminological database to associate texts with a specific code from the LOINC standard.
- Walter Koza's "Enumerative Series in Spanish: Formalization and Automatic Detection" presents a set of syntactic local grammars used to recognize and process enumerations in Spanish.
- Mohamed Aly Fall Seideh, Hela Fehri, and Kais Haddar's article "Recognition and Extraction of Latin Names of Plants for Matching Common Plant Named Entities" uses the International Code of Botanical Nomenclature as an interface to build an automatic French–Arabic term translation system.
- Hiba Chenny and Slim Mesfar's article "Generating Alerts from Automatically Extracted Tweets in Standard Arabic" presents a text parser capable of mining texts to extract relevant information from tweets written in Arabic.

The articles in the third part describe the construction of sophisticated syntactic grammars and the use of such grammars by automatic paraphrasing generators built with NooJ:

- Krešimir Šojat, Božo Bekavac, and Kristina Kocijan's article "Detection of Verb Frames with NooJ" presents a parser capable of automatically recognizing in Croatian texts derived verbs using morphological grammars, and then associate each recognized form with its syntactic valency, thanks to a chunker that uses a set of syntactic local grammars.
- Peter A. Machonis's article "Phrasal Verb Disambiguation Grammars: Cutting Out Noise Automatically" presents a set of English dictionaries and grammars used to automatically disambiguate sequences of texts that may (or may not) represent phrasal verb constructions.
- Mario Monteleone's article "NooJ Local Grammars for Endophora Resolution" presents the different types of endophora, and proposes several techniques using NooJ local grammars to solve them.
- Alberto Maria Langella's article "Paraphrases for the Italian Communication Predicates" presents an Italian transformational grammar capable of producing paraphrases of sentences that contain a communication predicate.
- Cristina Mota, Anabela Barreiro, Francisco Raposo, Ricardo Ribeiro, Sérgio Curto, and Luísa Coheur's article "eSPERTo's Paraphrastic Knowledge Applied to Question-Answering and Summarization" presents the eSPERTo paraphrastic engine and its use to produce summaries of Portuguese texts.

The articles in the last part of this volume describe business applications built with NooJ that use deep semantic analysis, using complex syntactic and semantic dictionaries and grammars.

- Maria Pia di Buono's article "Endpoint for Semantic Knowledge (ESK)" presents the ESK framework and shows how NooJ can be used to process semantic information using its dictionaries and grammars.
- Francesca Esposito and Maddalena della Volpe's article "Using Text Mining and Natural Language Processing to Support Business Decision: Towards a NooJ Application" presents an automatic decision-making system capable of parsing business documents using specialized linguistic resources.
- Héla Fehri, Mohamed Aly Fall Seideh, and Sondes Dardour's article "A Decision-Support Tool of Medicinal Plants Using NooJ Platform" presents an automatic system capable of recommending medicinal plants based on the recognition of various criteria present in a French text such as the age or symptoms.

This volume should be of interest to all users of the NooJ software because it presents the latest development of the software as well as its latest linguistic resources. To date, there are NooJ modules available for over 50 languages; more than 3,000 copies of NooJ are being downloaded each year.

Linguists as well as computational linguists who work on Arabic, Belarusian, Chinese, Croatian, English, French, Italian, Kabylian, Portuguese, Spanish, or Quechua will find advanced, up-to-the-minute linguistic studies for these languages in this volume.

We believe the reader will appreciate the importance of this volume, both for the intrinsic value of each linguistic formalization and the underlying methodology as well as for the potential for developing NLP applications along with linguistic-based corpus processors in the social sciences.

February 2017   ·   Linda Barone
                                                      Max Silberztein
                                                      Mario Monteleone

Organization

Scientific Committee

Xavier Blanco	Autonomous University of Barcelona, Spain
Yuras Hetsevich	United Institute of Informatics Problems, Belarus
Svetla Koeva	University of Sofia, Bulgaria
Peter Machonis	Florida International University, USA
Slim Mesfar	University of Manouba, Tunisia
Mario Monteleone	University of Salerno, Italy
Johanna Monti	University of Sassari, Italy
Karel Pala	Masaryk University, Czech Republic
Vladimír Petkevič	Charles University, Prague
Jan Radimský	University of South Bohemia, Czech Republic
Max Silberztein	Université de Franche-Comté, France
Marko Tadic	University of Zagreb, Croatia
François Trouilleux	Université Blaise-Pascal, France

Organizing Committee

Jan Radimský	University of South Bohemia, Czech Republic
Max Silberztein	Université de Franche-Comté, France
Zuzana Nevěřilová	Masaryk University, Czech Republic
Petr Kos	University of South Bohemia, Czech Republic
Mario Monteleone	Università degli studi di Salerno, Italy

Contents

Syntactic Analysis

Semantic Analysis and Its Applications

Vocabulary and Morphology

Treatments of the Kabylian Derived Nominal Verbs with NooJ

Hamid Annouz[✉]

INALCO, Paris, France
hamid_annuz@yahoo.fr

Abstract. In the following pages, we will present a part of a doctoral thesis which deals with the introduction of Kabylian language in a Natural Language Processing (NLP) platform called NooJ. We will try to show, mainly, how we have treated some derived nominal verbs, their inflectional patterns and the links to the verbs they come from. Afterwards, we will show some linguistic units from several dictionaries. Finally, we will apply the whole of our resources to a text, in order to show what results we obtain.

Keywords: Language automated treatment · NooJ application · Derivational · Kabylian verbs · Text annotation

1 Introduction

Kabylian dialect is a Berber language (from North Africa) belonging to the vast Afro-Asian language family. Most of the work done on Kabylian language and mainly on Berbere languages deal with the description of the language itself. However, Kabylian language benefits from a stable and generalised norm of writing from about 20 years, mainly in schools. As for the written language, it is taking a significant place.

Several aspects have allowed us to start the fore-mentioned project of introducing Kabylian language in the NLP field. For instance, the already cited breakthrough in the field of Kabylian language, the existence of dictionaries and books dealing with descriptive grammar, like for example the one by Naït-Zerrad (2001), and also the manual of conjugation by Naït-Zerrad (1994).

We are taking part in the fore-mentioned project to contribute and give some answers and outcomes in the field of linguistics of this new century, especially to reinforce the already strong link between technology and informatics. We work on this subject despite our consciousness about the limits and obstacles (already existing) we will encounter. However, as it has been done with other languages, we intend to introduce Kabylian in the field of NLP. We will be then able to modify, correct and add information. This is the very reason why we think that project is useful and possible.

We wish to show in this paper in what our work consists. We will follow these steps:

- Constitution of the inflectional models for the fore-mentioned derived words;
- In the dictionary, adding inflections to the derived words;
- Integration of prepositions, adverbs and affixes;

© Springer International Publishing AG 2016
L. Barone et al. (Eds.): NooJ 2016, CCIS 667, pp. 3–13, 2016.
DOI: 10.1007/978-3-319-55002-2_1

- Application to a text and results;
- Annotations;
- Problems;
- Prospects.

2 Treatment of an Example

Treatment of an example: this example constitute a dictionary entry to the finished state (a verbal entrance with all its derived forms).

acar,V+RV+FLX=AGAD+DRV=DERVMY1:MYAGAD+DRV=AČURAN:AZAY
AN+DRV=AČARAN:AZAYAN+DRV=TACART:TAFAT+DRV=INIG:ACAR

    The first derived form and its inflection +(DRV = DERVMY1:MYAGAD) is a verb which we will not take into consideration in this presentation.

3 Construction of the Model in the Grammar Part for the Verbal Derived Part of *Acar* (Fill)

3.1 Derivational Models

- AČURAN = *an* was added at the beginning and the letters *ca* was replaced with *ču* to obtain the adjective (A) *ačuran* (Fig. 1).

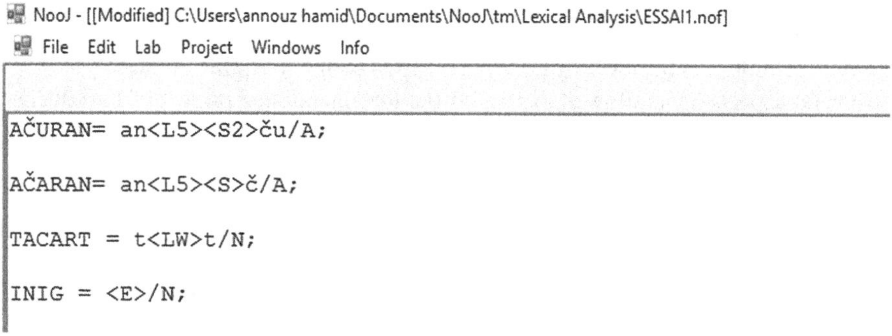

NooJ - [[Modified] C:\Users\annouz hamid\Documents\NooJ\tm\Lexical Analysis\ESSAI1.nof]
  File Edit Lab Project Windows Info

```
AČURAN= an<L5><S2>ču/A;

AČARAN= an<L5><S>č/A;

TACART = t<LW>t/N;

INIG = <E>/N;
```

Fig. 1. Example of derivational model

- AČARAN = = *an* was added at the beginning and the letter *c* with *č* to obtain the adjective *ačuran* (a variant of *ačuran*)
- TACART = the discontinuous morpheme *t—t* to have *tacart*, the result of the action for the verb *acar*.
- INIG = the name of the action *acar* has the same form of the verb *acar*. The verb and the name of the action are homographs.

3.2 Inflectional Models

Inflectional models describe the different forms already presented. For each form there is a derivational form presented in the models below (Fig. 2):

- TAFAT = inflectional paradigm of TACART. It is invariable and it has a feminine singular form.
- ACAR = inflectional paradigm of INIG. The form is masculine singular.
- AZAYAN = inflectional paradigms *AČURAN* and *AČARAN*. Their inflection is for the genre, number and status (free form and annexation form[1]).
- ačuran: masculine singular
  - | tačurant: feminine singular
  - | ičuranen: masculine plural
  - | tičuranin: feminine plural
  - | učuran: masculine singular (annexation form)
  - | tčurant: feminine singular (annexation form)
  - | yičuranen: masculine plural (annexation form)
  - | tčuranin: feminine plural (annexation form)

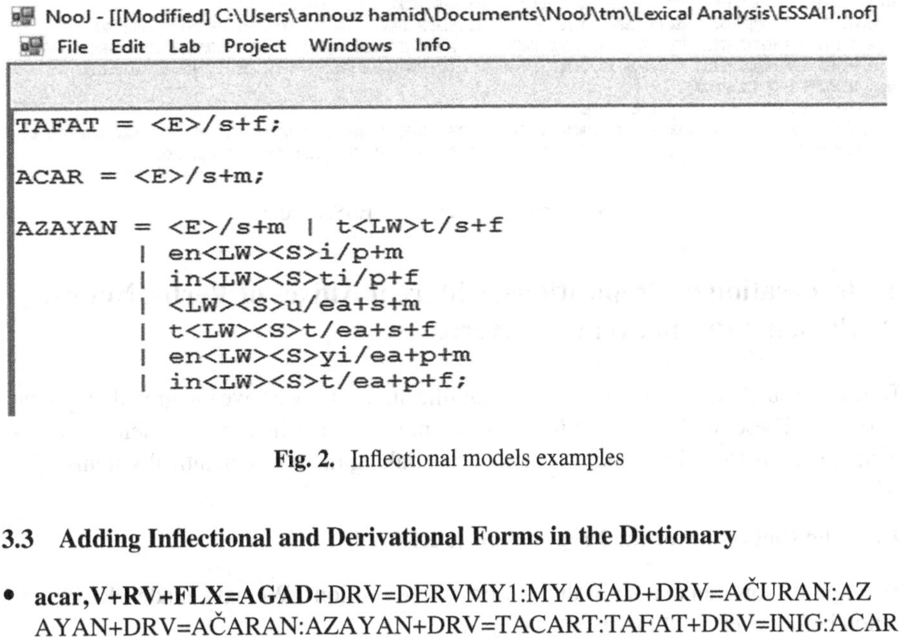

```
NooJ - [[Modified] C:\Users\annouz hamid\Documents\NooJ\tm\Lexical Analysis\ESSAI1.nof]
File   Edit   Lab   Project   Windows   Info

TAFAT = <E>/s+f;

ACAR = <E>/s+m;

AZAYAN = <E>/s+m | t<LW>t/s+f
         | en<LW><S>i/p+m
         | in<LW><S>ti/p+f
         | <LW><S>u/ea+s+m
         | t<LW><S>t/ea+s+f
         | en<LW><S>yi/ea+p+m
         | in<LW><S>t/ea+p+f;
```

Fig. 2. Inflectional models examples

3.3 Adding Inflectional and Derivational Forms in the Dictionary

- **acar,V+RV+FLX=AGAD+DRV=DERVMY1:MYAGAD+DRV=AČURAN:AZ AYAN+DRV=AČARAN:AZAYAN+DRV=TACART:TAFAT+DRV=INIG:ACAR**

The first line of the dictionary (acar) is a reversible verb with its inflectional model AGAD. This verb has its derivational form in MY with the inflectional model

---

[1] In Kabylian language, a word in "free form" means that it is in its initial form. A world in "annexation form" means that its first vowel was changed. This rule is present only in some syntactical context.

MYAGAD. The model AČURAN has the inflectional form AZAYAN. The model ACARAN has the inflectional form AZAYAN. The model TACART has the inflectional form TAFAT and the derivational form INIG. This form INIG has the inflectional form ACAR (Fig. 3).

```
###################################
#use Essail.nof

acar,V+RV+FLX=AGAD+DRV=DERVMY1:MYAGAD+DRV=AČURAN:AZAYAN+DRV=AČARAN:AZAYAN+DRV=TACART:TAFAT+DRV=INIG:ACAR
aḍen,V+I+FLX=AFEG+DRV=DERVS3:SSIWEḌ+DRV=DERVMS15:MSIREM+DRV=AMUḌIN:AMUDIN+DRV=ASIFEG:ASIMES+DRV=ATTAN:AṬṬAN
addi,V+RV+FLX=AGI+DRV=TADDUYT:TAFAT
adef,V+I+FLX=AFEG+DRV=DERVS3:SSIWEḌ+DRV=TUFSIN:TAFAT+DRV=ASIFEG:ASIMES
ader,V+I+FLX=AFEG+DRV=DERVS3:SSIWEḌ+DRV=DERVMS15:FRIWES+DRV=DERVN6:CRURED+DRV=TUFSIN:TAFAT+DRV=TACART:TAFAT
ader,V+T+FLX=AFEG+DRV=DERVTTW1:TTWAREZ+DRV=INIG:ACAR
adu,V+T+FLX=ARGU+DRV=AMDI:AMUDI
aɛer,V+I+FLX=AFEG##Remarque43
aɛlay,V+I+FLX=ALWAY+DRV=DERVS31:SSILWI+DRV=DERVS35:SBERWI+DRV=ALQAYAN:AMYAR+DRV=INIG:ACAR+DRV=FAD:ACAR+DRV=ASIFEG:ASIMES
aɛtad,V+FLX=AKLAL
aɛu,V+RV+FLX=ARU+DRV=DERVMY1:MFI+DRV=DERVMS12:MFI+DRV=TUGIN:TAFAT+DRV=AMYALI:AFSAR
aɛzag,V+I+FLX=ISHIL+DRV=INIG:ACAR
af,V+T+FLX=AF+DRV=DERVTTW1:TTUNEFK+DRV=DERVMY1:MYAF+DRV=TIFIN:TAFAT+DRV=AMYALI:AFSAR
afeg,V+I+FLX=AFEG+DRV=DERVS3:SSIWEḌ+DRV=DERVSR1:SRIFFEG+DRV=ASIFEG:ASIMES+DRV=AFFUG:AKAL+DRV=TIMSIN:TACART
afes,V+RV+FLX=AFEG+DRV=DERVTTW1:TTWAREZ+DRV=TUFSIN:TAFAT+DRV=ATTWACLEX:ACAR
aɣ,V+RV+FLX=AɣY+DRV=DERVMY1:MYAGAD+DRV=DERVMS12:MQALAC+DRV=DERVS3:SFAD+DRV=DERVTTW1:TTUNEFK+DRV=TIFIN:TAFAT
agad,V+I+FLX=AGAD+DRV=DERVS3:SSIWEḌ+DRV=DERVMY1:MYAGAD+DRV=ASIGGED:ASIMES+DRV=AMLAWI:AZAYAN+DRV=TIGDI:TAFAT+DRV=TIMSIN:TAFAT
aggad,V+I+FLX=AGAD+DRV=DERVS3:SSIWEḌ+DRV=DERVMY1:MYAGAD+DRV=ASIGGED:ASIMES+DRV=AMLAWI:AZAYAN+DRV=TIGDI:TAFAT+DRV=TIMSIN:TAFAT
agar,V+I+FLX=AGAD+DRV=DERVS3:SSIWEḌ+DRV=ASIGGED:ASIMES+DRV=TACART:TAFAT+DRV=TUGARIN:TAFAT+DRV=TASIGERT:TAFAT+DRV=TASAGERT:TAFAT
agat,V+I+FLX=AGAD+DRV=INIG:ACAR
agem,V+T+FLX=AFEG+DRV=ɛAWAZ:ACAR+DRV=ASALI:AZAYAN
aɡew,V+T+FLX=AFEG+DRV=DERVS4:JJIĜEW+DRV=DERVMY1:MDAFEɛ+DRV=DERVMY9:MSIREM+DRV=DERVMS9:MSIREM+DRV=ASIFEG:ASIMES+DRV=AZIĜEW:ASIMES+DRV=ɛ
agi,V+I+FLX=AGI+DRV=DERVMY1:MYAGI+DRV=DERVMS12:MYAGI+DRV=INIG:ACAR+DRV=TIGIN:TAFAT+DRV=TUGIN:TAFAT+DRV=AMYALI:AFSAR
```

Fig. 3. Part of the dictionary ESSAI1.dic

4   Integration of Prepositions, Different Affixes of Verbs, Nouns, Personal Pronouns and Adverbs

To enlarge our database and obtain more significant results, we have integrated linguistic resources. These dictionaries do not need too much effort whether at to their construction; the aim is to build up lists of dictionaries linking them to semantically items.

4.1   The Dictionary of the Prepositions (PREPA.Dic)

This figure shows an extract of the dictionary prepositions. We have made PREP indication before every dictionary entry (Fig. 4).

```
Dictionary contains 23 entries

# Special Characters: '\' '"' ' ' ',' '+' '-' '#'
#
# dic.prépa.dic

deg, PREP
di, PREP
d, PREP
yef, PREP
ddaw, PREP
seddaw, PREP
deffir, PREP
zdeffir, PREP
```

Fig. 4. Dictionary of prepositions (PREPA.dic)

4.2   The Dictionary of Verbs Affixes (AFFV.Dic)

On the first line there is the first lexical entry iyi that is an affix of the verb and it is the first singular masculine and feminine person (Fig. 5).

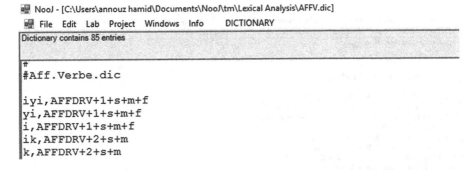

Fig. 5. Extract of the dictionary of verbs affixes

4.3   The Dictionary of the Possessive Affixes (POSS.Dic)

The first lexical entry presents in the Fig. 6, shows that iw is a possessive pronoun of the first singular masculine and feminine form.

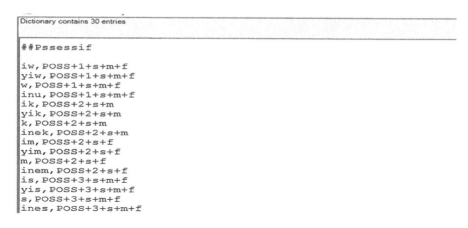

Fig. 6. Dictionary of possessives

4.4   The Dictionary of Personal Pronouns (PROPRS.Dic)

The first lexical entry of Fig. 7 nekk is an independent personal pronoun. It gives the first singular masculine and feminine form.

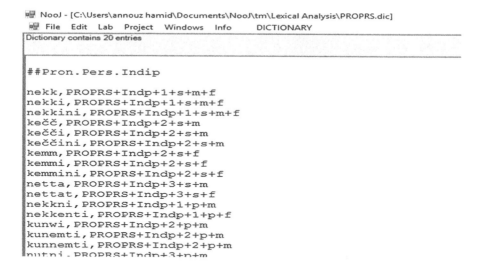

Fig. 7. Dictionary of independent personal pronouns

4.5   Dictionary of Adverbs (ADV.Dic)

The dictionary is composed of the simple forms ADV (adverb) without any specification (Fig. 8).

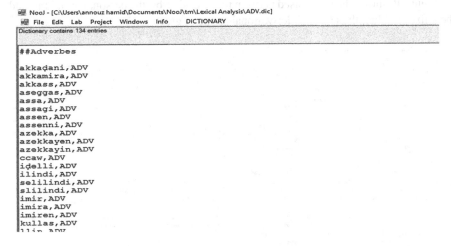

```
NooJ - [C:\Users\annouz hamid\Documents\NooJ\tm\Lexical Analysis\ADV.dic]
File   Edit   Lab   Project   Windows   Info      DICTIONARY
Dictionary contains 134 entries

"
##Adverbes

akkaḍani,ADV
akkamira,ADV
akkass,ADV
aseggas,ADV
assa,ADV
assagi,ADV
assen,ADV
assenni,ADV
azekka,ADV
azekkayen,ADV
azekkayin,ADV
ccaw,ADV
iḍelli,ADV
ilindi,ADV
selilindi,ADV
slilindi,ADV
imir,ADV
imira,ADV
imiren,ADV
kullas,ADV
llin  ADV
```

Fig. 8. Dictionary of adverbs

4.6   Dictionary of Prepositions Affixes (AFFPREP.Dic)

The first lexical entry in the Fig. 9 shows that *i* is a affixe of preposition for the first singular masculine and feminine form.

```
NooJ - [C:\Users\annouz hamid\Documents\NooJ\tm\Lexical Analysis\AFFPREP.dic]
File   Edit   Lab   Project   Windows   Info      DICTIONARY
Dictionary contains 28 entries

"
##Affixe.Prép

i,AFFPREP+1+s+m+f
ak,AFFPREP+2+s+m
k,AFFPREP+2+s+m
am,AFFPREP+2+s+f
m,AFFPREP+2+s+f
as,AFFPREP+3+s+m+f
s,AFFPREP+3+s+m+f
ney,AFFPREP+1+p+m
aney,AFFPREP+1+p+m
atney,AFFPREP+1+p+m
tney,AFFPREP+1+p+f
antey,AFFPREP+1+p+f
atentey,AFFPREP+1+p+f
wen,AFFPREP+2+p+m
awen,AFFPREP+2+p+m
atwen,AFFPREP+2+p+m
went,AFFPREP+2+p+f
awent,AFFPREP+2+p+f
atwent,AFFPREP+2+p+f
kent,AFFPREP+2+p+f
akent,AFFPREP+2+p+f
akwent,AFFPREP+2+p+f
```

Fig. 9. Dictionary of prepositions affixes

5   Application to a Text

The created linguistic resources have to be applied to a text to obtain the preliminary results.

Figure 10 shows the recognised verbs in red, the adverbs, the personal pronouns, and the affixes in blue, the nominal derivational forms in green.

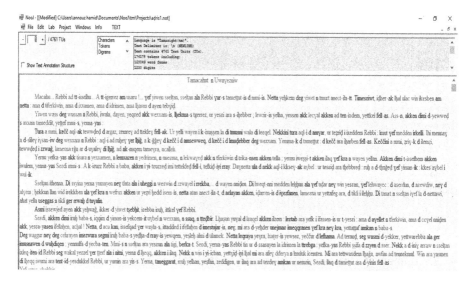

Fig. 10.   The recognised lexical forms in the text. (Color figure online)

5.1   The Annotations

See Fig. 11

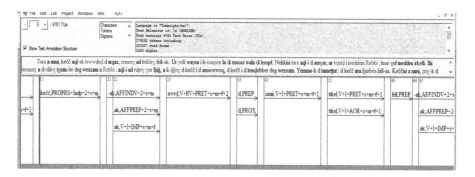

Fig. 11.   Annotations

6 Problems

6.1 The Silence

- *a*: vocative (not considered);
- *mmi*: primary noun (my son);
- *aql-*: presentative;
- *argaz*: primary noun (man);
- *ad:* preverbal particle (aoriste);

6.2 The Noise

- *tura*: verbal or adverbial form;
- *ak*: verbal form or preposition affixe or verb affixes (form direct);
- *d*: preposition or spatial particle;
- *ttkel*: verbal form of preterite or aoriste;

6.3 Syntactical Graphs (Non Verbal Forms)

Considering the problems already explained, a set of syntactical graphs are built and showed in the next sections.

6.4 The Affixes Followed by the Verbs (AFF-V.nog)

See Fig. 12.

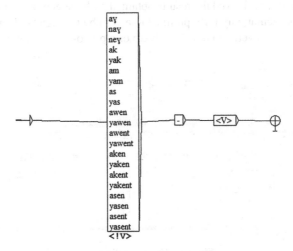

Fig. 12. The affixes followed by the verbs

6.5   The Affixes After the Verbs or Nouns or Prepositions or Adverbs and Followed by "-"

See Fig. 13

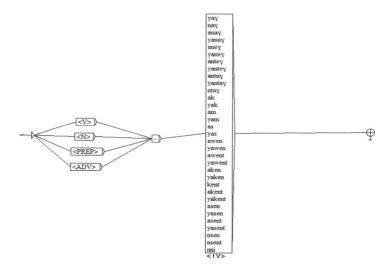

Fig. 13. The affixes after verbs, nouns, prepositions and adverbs.

7   Results

In these annotations are showed the results obtained with the application of this syntactical graphs. In particular, any form prefixed to a verb by the sign "-" is not a verb, and any suffix attached to a verb, a noun, an adverb or a preposition is not a verb (Fig. 14).

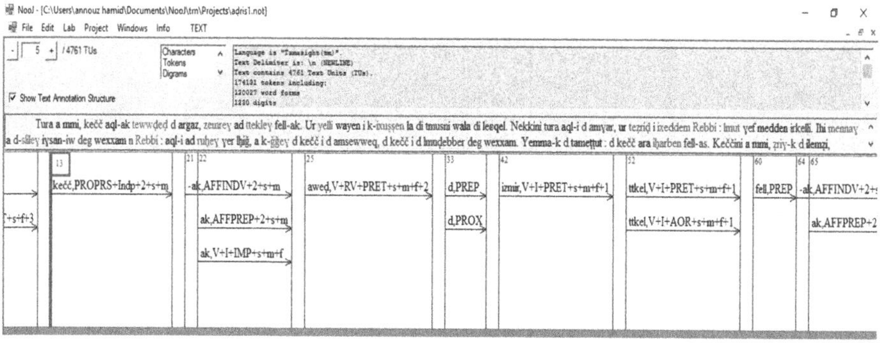

Fig. 14. The annotations.

8   Conclusions and Perspectives

In this presentation, Nooj linguistic tool is used to describe the nominal and derivational forms of the Kabylian language while maintaining the link with the base lexical entry. Considering this background, a set of syntactical graphs is built to solve the ambiguous forms (e.g.: affixes connected to the verbs, adverbial forms with verbal forms and so on). This project explains the most important problems to solve for improving the automatic processing of Kabylian language. The perspective aims to construct a dictionary of primary nouns and its derivational forms, a dictionary of the others lexical entries with their semantic and syntactic proprieties.

References

Bouillon, P., Vandooren, F., Sylva, L.D., Jacqmin, L., Lehmann, S., Russell, G., Viegas, E.: Traitement automatique des langues naturelles. Duculot, Paris (1998)

Fuchs, C.: Linguistique et traitement automatique des langues. Hachette, Paris (1993)

Dallet, J-M.: Dictionnaire Kabyle-Français, Parler des at Manguellat Algérie. SELAF, Paris (1982)

Dallet, J.-M.: II Dictionnaire Français-Kabyle. SELAF, Paris (1985)

Naït-Zerrad, K.: Grammaire moderne du kabyle, tajerrumt n teqbaylit. Karthala, Paris (2001)

Naït-Zerrad, K.: Manuel de conjugaison kabyle. L'Harmattan, Paris (1994)

Silberztein, M.: La formalisation des langues: l'approche de NooJ. ISTE Ed. Londres (2015)

Chaker, S.: Un parler berbère d'Algérie (Kabylie). Syntaxe, Thèse de troisième cycle, Université de Provence (1978)

Addition of IPA Transcription to the Belarusian NooJ Module

Stanislau Lysy[✉], Hanna Stanislavenka[✉], and Yury Hetsevich

The United Institute of Informatics Problems, National Academy of Sciences of Belarus,
Minsk, Belarus
stanislau.lysy@gmail.com, hanna.stanislavenka@gmail.com,
yury.hetsevich@gmail.com

Abstract. This paper is based on earlier research works where the possibility was shown to represent a linguistic phonetic level in NooJ. The phonetic level was developed for the Belarusian NooJ Module and was embodied with the help of dictionaries, with transcriptions in different formats. However, while Cyrillic transcriptions traditionally used for Belarusian were generated and compiled correctly, the international format had inaccuracies. Among these, one is connected with stress positions in IPA transcription. The task of adding IPA transcription requires the solving of problems concerning splitting words into syllables. Thus, firstly we need to solve the problem of syllabification of words in Belarusian.

Keywords: NooJ · Belarusian language · Phonetics · IPA transcription

1 Introduction

NooJ as a linguistic development environment supplies tools to describe all levels of natural language. The main sphere of our interest in this research paper is phonetics. In earlier research papers, we have shown that it is possible to represent a phonetic level of language in NooJ [1].

We demonstrated this new yet undeveloped possibility of NooJ in the Belarusian module by creating the dictionary with transcriptions and by developing morphological NooJ grammars with the help of which one can create phonetic transcriptions of arbitrary orthographic words. Results that were presented in NooJ 2015 Proceedings are the following [1]:

1. The system of generating of dictionaries in NooJ format with four types of phonetic transcriptions was created.
2. NooJ-dictionary was successfully compiled and tested. It contained 46.384 first forms of nouns in 2015.
3. Development of morphological NooJ grammar for creating a phonetic transcription for orthographic words was launched.

However, there are still some issues in generating of transcriptions. One of them is connected with stress position in some types of transcriptions (in particular IPA

© Springer International Publishing AG 2016
L. Barone et al. (Eds.): NooJ 2016, CCIS 667, pp. 14–22, 2016.
DOI: 10.1007/978-3-319-55002-2_2

transcription), where stresses (') are put before the stressed syllables in the phonetic word transcription. The solution of this issue requires multitasking work that we split into three main steps. Syllabification algorithm for generation of IPA transcriptions of orthographic words in Belarusian should be developed. Then it would be possible to advance high-quality tool for the generation of phonetic transcription of orthographic words in Belarusian [2]. With the help of the advanced tool, a dictionary in NooJ format for nouns and verbs can be created.

2 IPA-Transcription in Belarusian NooJ-Module

Above mentioned NooJ-dictionaries contains the following types of phonetic transcription:

- Cyrillic transcription (based on [3]);
- International Phonetic Alphabet (or IPA) [4];
- Simplified international format [5];
- Extended Speech Assessment Methods Phonetic Alphabet (or X-SAMPA) [6].

This paper is dedicated to the IPA as it is used worldwide by lexicographers, foreign language students and teachers, linguists, speech-language pathologists, other researchers, as well as singers, actors, anchors. Available NooJ dictionaries with international transcription format will contribute to the development and formalizing of the Belarusian language.

There are various traditions of marking stresses in a word transcription. Thus a stress mark in a word can be connected with a certain allocation of a stressed vowel sound (a stress mark above or after vowel, vowel written in uppercase, and so on), or with an allocation of a stressed syllable (a stress mark before syllable, after syllable, writing syllable in uppercase, and so on). The IPA-transcription marks stress not on the vowel (as in tradition for the Belarusian language) but before an accented syllable. The system of NooJ dictionaries generation puts a stress mark incorrectly for the IPA-format [2]:

смертнасць,NOUN+TranscriptionIPA=[sʲmʲˠɛrtnasʲtsʲ]

The correct variant is as follows:

смертнасць,NOUN+TranscriptionIPA=[ˠsʲmʲɛrtnasʲtsʲ]

Such seemingly irrelevant incorrectness can lead to misconstrued usage of the transcription of Belarusian words and correspondingly to the wrong results of research, and so on. A solution of the problem lies in developing of syllabification algorithm which is for the first time presented in the next part of paper.

3 Syllabification Process

As it was mentioned above syllabification process is firstly demonstrated as a computational linguistic problem for the Belarusian language in this paper. For the algorithm syllabification rules were developed according to the book "Phonetics of the Belarusian Literary Language" [3]. Developed rules have the following view:

'aa' => 'ala'
'aka' => 'alka'
'ama' => 'alma'
'akka' => 'alkka'
'akma' => 'alkma'
'amka' => 'amlka'
'amma' => 'amlma'
'akkka' => 'alkkka'
'akkma' => 'alkkma'
'akmka' => 'akmlka'
...

where "a" is an arbitrary vowel, "k" is an obstruent consonantal, "m" is a sonorant consonantal, and "l" stands for a syllable border.

For instance, the rule "'akmka' => 'akmlka'" means that if between two vowels there are three consonant phonemes: one is obstruent, one is sonorant and one is again obstruent, first two phonemes will be in one syllable, and syllable border must be put before the second obstruent phoneme.

The algorithm works with certain compulsory data and has 5 main steps. Input data for the algorithm is:

– arbitrary word in Belarusian in an allophonic format W_a;
– set of syllabification rules $R_{syll} = \ll Pt_1, Pt_{syll1} > ,..., < Pt_m, Pt_{syllm} \gg$, where Pt_i – sequence of allophonic class notations in the ith rule, Pt_{sylli} – sequence of allophonic class notations with a syllable border in the ith rule, $i = 1...m$, m –number of syllabification rules;
– set of correspondences "allophone – class of allophone", $A_{class} = \ll A_1, Cl_1 > ,$ ..., $< A_n, Cl_n \gg$, where A_i – allophone code, Cl_i – allophone class notation, $i = 1 ...n$, n – number of allophones.

Note that in order to determine a position of a syllable border allophones were divided into three classes: vowel (notation "a"), obstruent consonantal (notation "k"), and sonorant consonantal (notation "m"). Scheme of the algorithm (Fig. 1) and description of its work steps are presented below:

Step 1. An input word W_a is divided into allophones, that are consecutively put into a list $L_a = < A_{w1},..., A_{wn} >$, where A_{wi} – the ith allophone in a word, $i = 1...n$, n – number of allophones in a word.

Step 2. Sequence of allophone class notations is formed from the list L_a. For each allophone A_{wi} from the list L_a there is an allophone class notation Cl_i in the list A_{class}. Found notations are collected in a word pattern according to the classes of its allophones Pt_w.

Step 3. Consecutive browsing of Pt_i pattern from the set of syllabification rules R_{syll} is held. Algorithm searches each Pt_i pattern in a pattern of the input word template Pt_w. When occurrence of the Pt_i pattern is found in the input word pattern Pt_w, Pt_i pattern is replaced with relative pattern but with a syllable border Pt_{sylli} from the set of rules R_{syll} in the input word pattern Pt_w.

Step 4. The algorithm is browsing symbols of the modified Pt_w pattern. If i^{th} symbol of the Pt_w pattern is a syllable border, then the ith element with a syllable border is added to a list if allophones L_a of the input word W_a.

Step 5. The modified allophone list L_a of the input word W_a is elementwise assembled into one allophonic word with syllable borders W_{ares}.

Step 6. End of algorithm. The result of the algorithm is a syllabified allophonic word.

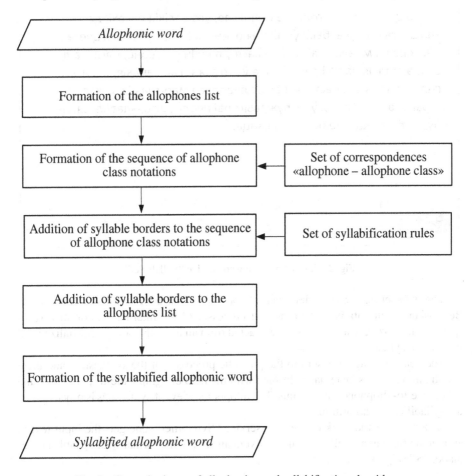

Fig. 1. General scheme of allophonic word syllabification algorithm

4   Software Prototype "Syllabifier"

For testing above described algorithm, a web-service "Syllabifier" was created. "Syllabifier" is a program prototype that embodies the allophonic word syllabification algorithm [7]. Its interface is presented in Fig. 2.

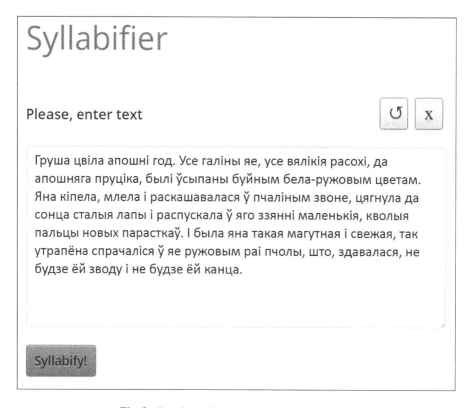

Fig. 2. Interface of the web-service "Syllabifier"

The web-service "Syllabifier" can take as input data both word lists and texts in Belarusian. Input data is sent to the text processor of the Internet-version of text-to-speech synthesizer, where words are extracted from an input list or text, normalized and united into syntagmas [8].

Then each syntagma is sent to the phonetic processor of the synthesizer and there allophonic view of syntagmas is formed with the help of "grapheme-to-phoneme" and "phoneme-to-allophone" algorithms. The allophonic view of syntagmas is processed by the syllabification algorithm.

As a result of the work of the web-service "Syllabifier", one gets the input text in allophonic format, as well as the input text in an allophonic format with syllable border marks ">" (Fig. 3).

Text in allophonic format

GH004,R022,U022,SH002,A323,/,C'002,V'002,I241,L002,A012,/,A221,P001,
O012,SH002,N'004,I242,/,GH001,O032,T000,/,#P4,
U203,S'001,E042,/,GH004,A233,L'002,I042,N004,Y323,/,J'012,A243,J'011,E
040,/,#C3,
U203,S'001,E043,/,V'012,A243,L'002,I043,K'002,I343,J'012,A342,/,R002,A2
22,S001,O023,H'002,I340,/,#C3,
D002,A022,/,A221,P001,O012,SH002,N'004,A342,GH004,A231,/,P002,R012
,U023,C'002,I342,K004,A330,/,#C3,
B002,Y013,L'004,I241,/,W013,S001,Y021,P002,A312,N004,Y321,/,B004,U21
3,J'013,N002,Y021,M001,/,B'002,E141,L004,A312,R002,U222,ZH002,O021,
V012,Y211,M003,/,C'002,V'001,E042,T002,A321,M000,/,#P4,
J'002,A242,N002,A023,/,K'002,I243,P'001,E041,L004,A310,/,#C3,
M001,L'002,E041,L004,A213,/,I022,/,R002,A322,S002,K004,A332,SH002,A2
21,V011,A011,L004,A313,S'002,A341,/,W013,/,P002,CH002,A223,L'002,I04

Syllabified text in allophonic format

>,GH004,R022,U022,>,SH002,A323,/,>,C'002,V'002,I241,>,L002,A012,/,>,A2
21,>,P001,O012,>,SH002,N'004,I242,/,>,GH001,O032,T000,/,>,#P4,
>,U203,>,S'001,E042,/,>,GH004,A233,>,L'002,I042,>,N004,Y323,/,>,J'012,A2
43,>,J'011,E040,/,>,#C3,
>,U203,>,S'001,E043,/,>,V'012,A243,>,L'002,I043,>,K'002,I343,>,J'012,A342,
/,>,R002,A222,>,S001,O023,>,H'002,I340,/,>,#C3,
>,D002,A022,/,>,A221,>,P001,O012,>,SH002,N'004,A342,>,GH004,A231,/,>,
P002,R012,U023,>,C'002,I342,>,K004,A330,/,>,#C3,
>,B002,Y013,>,L'004,I241,/,>,W013,S001,Y021,>,P002,A312,>,N004,Y321,/,>,
B004,U213,J'013,>,N002,Y021,M001,/,>,B'002,E141,>,L004,A312,>,R002,U2
22,>,ZH002,O021,>,V012,Y211,M003,/,>,C'002,V'001,E042,>,T002,A321,M00
0,/,>,#P4,
>,J'002,A242,>,N002,A023,/,>,K'002,I243,>,P'001,E041,>,L004,A310,/,>,#C3,

Fig. 3. Interface of web-service "Syllabifier" with output data

5   NooJ-Dictionaries Compilation

The described above algorithm was inbuilt to online service "Orthoepic Dictionary Generator" in which there is an option – "Headwords processing in NooJ format" [2]. It means that the first word in every line of the input text is processed and transcription of the word written in the NooJ format appears. After processing we get material with which we create Dictionary for NooJ. Here the fragment of generated NooJ dictionary is shown:

смертнасць,NOUN+TranscriptionCyr=[c'м'эֻртнас'ц']+TranscriptionIPA=[ˈsʲmʲɛrtnasʲʦʲ]
смертнік,NOUN+TranscriptionCyr=[c'м'эֻртн'ік]+TranscriptionIPA=[ˈsʲmʲɛrtnʲik]
смертніца,NOUN+TranscriptionCyr=[c'м'эֻртн'іца]+TranscriptionIPA=[ˈsʲmʲɛrtnʲiʦa]
смертухна,NOUN+TranscriptionCyr=[c'м'эֻртухна]+TranscriptionIPA=[ˈsʲmʲɛrtuxna]
смерць,NOUN+TranscriptionCyr=[c'м'эֻрц']+TranscriptionIPA=[ˈsʲmʲɛrʦʲ]
смерч,NOUN+TranscriptionCyr=[c'м'эֻрч]+TranscriptionIPA=[ˈsʲmʲɛrʧ]
сметнік,NOUN+TranscriptionCyr=[c'м'эֻтн'ік]+TranscriptionIPA=[ˈsʲmʲɛtnʲik]
сметніца,NOUN+TranscriptionCyr=[c'м'эֻтн'іца]+TranscriptionIPA=[ˈsʲmʲɛtnʲiʦa]
сметнішча,NOUN+TranscriptionCyr=[c'м'эֻтн'ішча]+TranscriptionIPA=[ˈsʲmʲɛtnʲiʂʧa]
смех,NOUN+TranscriptionCyr=[c'м'эֻх]+TranscriptionIPA=[ˈsʲmʲɛx]
смехата,NOUN+TranscriptionCyr=[c'м'эхатаֻ]+TranscriptionIPA=[sʲmʲɛxaˈta]
смешкі,NOUN+TranscriptionCyr=[c'м'эֻшк'і]+TranscriptionIPA=[ˈsʲmʲɛʂkʲi]
смірна,NOUN+TranscriptionCyr=[c'м'іֻрна]+TranscriptionIPA=[ˈsʲmʲirna]
смог,NOUN+TranscriptionCyr=[смоֻх]+TranscriptionIPA=[ˈsmɔx]
смок,NOUN+TranscriptionCyr=[смоֻк]+TranscriptionIPA=[ˈsmɔk]
смоква,NOUN+TranscriptionCyr=[смоֻква]+TranscriptionIPA=[ˈsmɔkva]

As it is seen on the fragment above, stresses in IPA transcriptions are put in correct places.

With the help of the web-service "Orphoepic Dictionary Generator" we can compile dictionaries for NooJ with many entries. Thus we made a dictionary with nouns and with verbs [2]. A dictionary for nouns contains more than 49 000 entries An extract from this NooJ dictionary one can see on Fig. 4.

Entry	Category	TranscriptionCyr	TranscriptionIPA
смертнасць	NOUN	[c'м'эֻртнас'ц']	[ˈsʲmʲɛrtnasʲʦʲ]
смертнік	NOUN	[c'м'эֻртн'ік]	[ˈsʲmʲɛrtnʲik]
смертніца	NOUN	[c'м'эֻртн'іца]	[ˈsʲmʲɛrtnʲiʦa]
смертухна	NOUN	[c'м'эֻртухна]	[ˈsʲmʲɛrtuxna]
смерць	NOUN	[c'м'эֻрц']	[ˈsʲmʲɛrʦʲ]
смерч	NOUN	[c'м'эֻрч]	[ˈsʲmʲɛrʧ]
сметнік	NOUN	[c'м'эֻтн'ік]	[ˈsʲmʲɛtnʲik]
сметніца	NOUN	[c'м'эֻтн'іца]	[ˈsʲmʲɛtnʲiʦa]
сметнішча	NOUN	[c'м'эֻтн'ішча]	[ˈsʲmʲɛtnʲiʂʧa]
смех	NOUN	[c'м'эֻх]	[ˈsʲmʲɛx]
смехата	NOUN	[c'м'эхатаֻ]	[sʲmʲɛxaˈta]
смешкі	NOUN	[c'м'эֻшк'і]	[ˈsʲmʲɛʂkʲi]
смірна	NOUN	[c'м'іֻрна]	[ˈsʲmʲirna]
смог	NOUN	[смоֻх]	[ˈsmɔx]
смок	NOUN	[смоֻк]	[ˈsmɔk]
смоква	NOUN	[смоֻква]	[ˈsmɔkva]

Fig. 4. Extract from the NooJ dictionary for nouns

Moreover, a dictionary with transcriptions for verbs was compiled. It contains more than 30 000 entries. An extract from this NooJ dictionary one can see in Fig. 5. With the help of the IPA transcription, anyone can read a word in Belarusian.

Entry	Category	TranscriptionCyr	TranscriptionIPA
дзынкаць	VERB	[z'ынкац']	['d͡zʲɨnkatsʲ]
дзынкнуць	VERB	[z'ынкнуц']	['d͡zʲɨnknutsʲ]
дзьмухаць	VERB	[z'мухац']	['d͡zʲmuxatsʲ]
дзьмухнуць	VERB	[z'мухнуц']	['d͡zʲmuxnutsʲ]
дзьмуцца	VERB	[z'муц:а]	['d͡zʲmut͡sːa]
дзьмуць	VERB	[z'муц']	['d͡zʲmutsʲ]
дзюбануць	VERB	[z'убануц']	[d͡zʲuba'nutsʲ]
дзюбацца	VERB	[z'убац:а]	['d͡zʲubat͡sːa]
дзюбаць	VERB	[z'убац']	['d͡zʲubatsʲ]
дзюбнуцца	VERB	[z'убнуц:а]	['d͡zʲubnut͡sːa]
дзюбнуць	VERB	[z'убнуц']	['d͡zʲubnutsʲ]
дзюрчаць	VERB	[z'урчац']	[d͡zʲur'tʃatsʲ]
дзюрчэць	VERB	[z'урчэц']	[d͡zʲur'tʃetsʲ]
дзявацца	VERB	[z'авац:а]	[d͡zʲa'vat͡sːa]
дзяваць	VERB	[z'авац']	[d͡zʲa'vatsʲ]
дзяжурыць	VERB	[z'ажурыц']	[d͡zʲa'ʐuritsʲ]
дзякаваць	VERB	[z'акавац']	['d͡zʲakavatsʲ]
дзяліцца	VERB	[z'ал'іц:а]	[d͡zʲa'lʲit͡sːa]
дзяліць	VERB	[z'ал'іц']	[d͡zʲa'lʲitsʲ]
дзяржацца	VERB	[z'аржац:а]	[d͡zʲar'ʐat͡sːa]
дзяржаць	VERB	[z'аржац']	[d͡zʲar'ʐatsʲ]
дзяўбацца	VERB	[z'аўбац:а]	[d͡zʲaw'bat͡sːa]
дзяўбаць	VERB	[z'аўбац']	[d͡zʲaw'batsʲ]
дзяўбці	VERB	[z'аўпц'і]	[d͡zʲaw'ptsʲi]
дзяўбціся	VERB	[z'аўпц'іс'а]	[d͡zʲaw'ptsʲisʲa]

Fig. 5. Extract from the NooJ dictionary for verbs

6   Conclusion and Further Steps to Take

Summarizing our research the following results should be underlined. Firstly, syllabification algorithm for generation of IPA transcriptions of orthographic words in Belarusian was developed and tested on a special tool. Secondly, high-quality tool for generation of phonetic transcription of orthographic words in Belarusian that was presented in the results of the previous research was advanced with the help of the developed syllabification algorithm. Thirdly, dictionaries in NooJ format that include correct phonetic transcriptions in IPA format for nouns and verbs were created.

The results presented in the paper are a part of the ongoing research. Provided that plans are to examine correctness of the IPA transcriptions for NOUN and VERB; to build NooJ morphology grammar for letter-to-phoneme conversion with right syllable positions; to add IPA-transcription for Adjectives and Adverbs NooJ dictionaries.

The results of this research can be used in introducing and learning the norms of the literary pronunciation of the Belarusian language. The process of syllabification can also be further implemented into other computational linguistic programs and into NooJ modules of other languages.

References

1. Zahariev, V., Lysy, S., Hiuntar, A., Hetsevich, Y.: Grapheme-to-phoneme and phoneme-to-grapheme conversion in belarusian with NooJ for TTS and STT systems. In: Okrut, T., Hetsevich, Y., Silberztein, M., Stanislavenka, H. (eds.) NooJ 2015. CCIS, vol. 607, pp. 137–150. Springer, Heidelberg (2016). doi:10.1007/978-3-319-42471-2_12
2. Orthoepic Dictionary Generator. Available at http://corpus.by/OrthoepicDictionaryGenerator (2016)
3. Падлужны, А.І.: Фанетыка беларускай літаратурнай мовы. Навука і тэхніка, Мінск (1989)
4. The International Phonetic Alphabet and the IPA Chart. Available at http://www.internationalphoneticassociation.org/content/ipa-chart (2016)
5. Кошчанка, У.А.: Беларуска-англійскі размоўнік. Артыя Груп, Мінск (2010)
6. Computer-coding the IPA: a proposed extension of SAMPA (2016). http://www.phon.ucl.ac.uk/home/sampa/x-sampa.htm
7. Syllabifier (2016). http://corpus.by/Syllabifier
8. Text-to-Speech Synthesizer (2016). http://corpus.by/TextToSpeechSynthesizer

Recognizing Diminutive and Augmentative Croatian Nouns

Kristina Kocijan[✉], Marijana Janjić, and Sara Librenjak

Department of Information and Communication Sciences,
Faculty of Humanities and Social Sciences, University of Zagreb, Zagreb, Croatia
{krkocijan,marijanji,slibrenj}@ffzg.hr

Abstract. In this paper, the authors present NooJ morphological grammars for recognizing Croatian diminutive and augmentative nouns for those common nouns that already exist in the Croatian NooJ dictionary. The purpose of this project is twofold. The first one is to recognize both diminutive and augmentative forms of each noun existing in our dictionary (over 20 000 common nouns) if such a form occurs in a text. The second purpose is to determine types of texts in which these words appear the most (or if they even appear) which is the reason why we divided our corpus in two thematic categories (children literature, novels). The results of our algorithm are high on both types of text [overall P = 0.82; R = 0.80; f-measure = 0.81]. Although NooJ dictionary allows direct entrance of such derivations as an attribute-value description of a main noun, we have opted for the second option, i.e. writing a morphological grammar that will recognize the needed form. In this way, we are saving the space and time needed to add all the existing forms to the noun's dictionary.

Keywords: Augmentatives · Diminutives · Croatian · Morphology · Nouns · Nooj

1 Introduction

Diminutives and augmentatives are a fruitful field of study as can be seen from the number of approaches taken up by authors in their research across languages: phonetics [10], morphology [5, 6, 8], semantic aspects [7, 13, 15] to name but a few. The most relevant work for this paper is the research in Croatian language, as our paper aims to add some new information and insight in diminutives and augmentatives in Croatian. The search for such research did not give us many results, but it was diverse and rich in information that helped us in our preliminary steps.

The authors of Croatian grammars and older generation of researchers [1, 2] focused mainly on the morphological issues in diminutives and augmentatives in general, offering a short overview of the dominant morphological patterns. On the other hand, another approach is taken in some more recent research papers. Thus, analysis of diminutive Croatian verbs was discussed in details in [9], while [4, 16] provide a more semantic and pragmatic view to the topic of Croatian diminutives. Some comparisons between diminutive and/or augmentative suffixes of Croatian and other languages are found in [15, 16, 19] among others.

© Springer International Publishing AG 2016
L. Barone et al. (Eds.): NooJ 2016, CCIS 667, pp. 23–36, 2016.
DOI: 10.1007/978-3-319-55002-2_3

The overview of the data shows that the definitions of diminutives and augmentatives are an issue on its own. Bosanac et al. [4] speculate that the lack of precise definitions in grammar might be caused by the lack of field research on the use of such lexemes. Thus, their paper points out to a great need for pragmatic scholarly research that either involves questionnaires, as is the case of research conducted by [4], or the corpus based research, as is the case in this paper.

We believe that this paper will be a small contribution to a study oriented towards the language usage with its two aims. The first one is to recognize both diminutive and augmentative forms of each noun existing in the Croatian NooJ dictionary (over 20 000 common nouns [20]) if such a form occurs in a text. At this point verbal and adjective forms are not taken into account as that would make the task very complex. Hence, we agreed to focus on a segment of a puzzle in order to test the methodology.

The reason for the variation in the corpus is the second question that concerns us: in which types of texts diminutive and augmentative nouns mostly appear. The reasons behind the combination of corpus sources is not just that the earlier scholarly work was mostly theoretical, but also that it did not always clearly state how certain results and statistics were gained. Thus, [1, 2] discuss the productivity of particular suffixes in nominal diminutives but does not offer information on the corpus this statistics is based on. In [1] it remains unclear why the year 1860 was relevant for the author's research on contemporary Croatian language, or what is the information on some of the most productive suffixes based on, as the author offers an explanation of 'general language acquaintance' as a relevant one. Such issues point out the importance of critical analysis of previous research in order to offer some new insight in the field of (Croatian) diminutives and augmentatives in general.

The first step as a step towards attaining the new insight is to build morphological grammars that will be able to recognize diminutive and augmentative occurrences regardless the number and the case of a noun. Similar work has already been reported for Serbian [11] and Portuguese [14]. Once built, the grammars would be able to recognize and classify the newly found nouns either as a diminutive or augmentative form of a noun existing in the NooJ dictionary [20]. For example, if there is a noun *kuća* (house) in the dictionary, the grammar will also recognize its diminutive form *kućica* (en. small house) and mark it as <N+DIM+MAIN=kuća+Case+Gender+Number>. The augmentative form *kućerina* (en. big house) will be marked after the same pattern as <N+AUG+MAIN=kuća+Case+Gender+Number>. The only prerequisite is that the main noun exists in the NooJ dictionary. In this paper, we will present how we built the morphological grammar, what issues have we encountered and how we propose to solve them. The conclusions that we offer are based on the experience we gained from building grammars and corpus analysis. An interested reader should however bear in mind that this is only the first phase of the research and hence, not all answers are available at the moment.

The remaining of this paper is structured in the following manner. We will start with providing some theoretical approaches that exist for Croatian diminutive and augmentative nouns and then will turn to the digital dictionary of nouns that we used as the basic platform for our grammars. We will proceed with an insight into the corpus we have used for this research that will be followed by the description of our grammars for detecting and annotating derivational forms. At the end of the paper, we will give the results we have collected so far and will provide some final thoughts on our results and the usability of our approach.

2   The Theory Behind Diminutives and Augmentatives in Croatian Language

When we talk about diminutives and augmentatives, we are talking about nouns that are much smaller (diminutives) or much bigger (augmentatives) in size or value or intensity than the average [4].

Contexts in which diminutive forms appear are mostly observed in **child directed speech** (*Daj mi lopticu[1]!* - en: Give me [little] ball), **teasing** (*Ti si moja slatka loptica[2]*. - en. You are my sweet [little] ball.), **affectionate utterance** (*Da vidimo taj vaš trbuščić[3]*. - en. Let us see that [little] belly of yours.) and in **jargon** (*Bokić[4]*. - en. [Little] hello.). Augmentative forms have been detected when **making fun of somebody or something** (*Koja glavurina.* - en. What a [huge] head.), when **insulting someone** – (*Ona je kravetina.* - en. She is a [huge] cow.; *On je konjina!* - en. He is a [huge] horse!), when **exaggerating** (*Koja zgradurina.* - en. What a [huge] building.) or for **exaltation of a positive characteristic** (*On je prava ljudina[5]*. - en. He is a [very big] person.). However, some times, depending on the context, some diminutive nouns may take augmentative meaning (ex. *psihić[6]*) and vice versa (ex. *glavonja[7]*) [16]. More on the context and the meanings of Croatian diminutives may be found in [4].

The list of suffixes used to build diminutive and augmentative forms is a close set. Some suffixes are more productive and are used for several hundred of nouns while some are used for only one or several nouns. The lists are gender dependent i.e. some suffixes are characteristic for only masculine (*-ić* | *-čić* | *-(a)k* | *-eč(a)k* | *-ič(a)k*), or only feminine (*-ca* | *-ica* | *-čica*), or only neutral nouns (*-ce* | *-ance* | *-ašce* | *-ence* | *-ešce*). Still, there are suffixes like *-eljak* and *-uljak* for diminutives that may be used for any gender nouns. Suffixes for augmentative nouns are *-ina* | *-čina* | *-etina* | *-urina* | *-erina* | *-ešina* | *-ura* | *-urda* | *-ušina* | *-uština* | *-eskara* | *-uskara* regardless the noun's gender. As in the case of some diminutives, a gender can be shifted. Consequently, masculine nouns that take suffixes *-ina, -čina, -etina, -urina, -usina,* or *-čuga* may either remain masculine nouns or may become feminine nouns.

In Croatian language we can produce diminutive and/or augmentative forms for the large number of nouns (both common and proper). In the course of work on NooJ grammar we have realized, however, that certain limits do exist and that there are nouns

---

[1] Expression is used regardless the size of a ball.
[2] Expression used for an overweight person.
[3] Expression used for a pregnant lady regardless the size of her belly.
[4] Greeting characteristic for the area of City of Zagreb.
[5] Expression used for a person of a big heart.
[6] Diminutive form of a noun 'psychiatrist' when talking about a great psychiatrist.
[7] Augmentative form of a noun 'head' when talking about someone who is always proposing some not so good ideas.

which do not produce diminutive and/or augmentative forms. Such are, for example, abstract nouns like *milosrđe* (en. mercy) or collective nouns like *cvijeće* (en. flowers), *suđe* (en. dishes), *unučad* (en. grandchildren), *otočje* (en. islands) and so on, which do not form either diminutives or augmentatives.

Most of the common neutral nouns that end in *-nje* are considered to be verbal nouns (i.e. they were built from verbs) and they do not make diminutive or augmentative forms either. The same applies to some other neutral nouns ending in *-nje* like *bezakonje* (en. lawlessness) or *trnje* (en. thorns), but there are some that are exceptions to this rule like *janje* (en. lamb). Majority of common neutral nouns ending in *-ce* (*lice* – en. face), *-če* (*unuče* – en. grandchild), *-će* (*proljeće* – en. spring), *-đe* (*suđe* – en. dishes), *-je*[8] (*otočje* – en. islands), *-lje* (*bilje* – en. plants), *-šte* (*stubište* – en. stairway), *-vo* (*pecivo* – en. pastry), *-stvo* (*sudstvo* – en. judiciary), *-štvo* (*divljaštvo* – en. savagery), *-ost* (*jednostavnost* – en. simplicity) do not have their diminutive or augmentative forms either.

At this stage we also became aware of the nouns which appear to be diminutives but they have ceased to be perceived that way by Croatian speakers, or they co-exist in both diminutive and non-diminutive meanings. Such is the case with the following examples: *vreća* (en.bag) – *vrećica* (en. little bag but also grocery bag), *vrt* (en. garden) – *vrtić* (en. kindergarten), *ploča* (en. panel) – *pločica* (en. little panel but also bathroom tile).

Next to the nouns that do not have diminutive forms, Babić's work [1–3] also made us aware of the nouns that produce several diminutive forms but with different meanings (ex. *kuća* (en. house): *kućica* – little house and *kućerak* – little poor house [1]). Such words present a specific semantic problem that falls outside the scope of this paper and thus will not be discussed here in more details.

There is also a class of nouns that appears to bear one of the augmentative morphemes, but actually only has a visual resemblance to augmentatives while their phonological elements differ [12]. For example, noun *pile* (en. chicken) when added suffixes *-ina* (which is an augmentative suffixes) builds a noun *piletina* with the meaning 'chicken meat' and not 'a [huge] chicken'. Another issue related to diminutives is their relation with hypocorisms in Croatian language, while augmentatives are also interrelated with pejoratives.

In the next section, we will give a short account of the Croatian nouns in NooJ dictionary before we turn to the detailed description of morphological grammars for detection of diminutives and augmentatives, that we have built for this project.

3   Dictionary of Croatian Nouns

Presently, there are 20 350 common nouns in Croatian NooJ Dictionary, of which 8 416 are feminine, 6 387 are masculine and 5 547 are of neutral gender [20]. However, 4 423 neutral nouns do not form either diminutive or augmentative forms. We can recognize this type of nouns by their type and gender information in addition to their endings. We use this information to seclude them from our morphological grammar recognition.

---

[8] However, *jaje* (en. egg) is an exception.

Since NooJ [18] allows the derivational paradigms to be added directly to the dictionary, one of the possible approaches to recognize diminutive and augmentative forms is to add +DRV attribute to each of the existing nouns for each derivation that is possible. Thus, while some of the nouns have only one derivation for either diminutive or augmentative form, some have both derivations, some neither and some may have several diminutive and/or augmentative derivations. Examples for the last type are found in [2] *kamen -> kamenčina, kamenčuga, kamenina*[9] (en. stone -> big stone), *noga -> nogetina, nožetina* (en. leg -> big leg)[10]. There are also nouns that share the diminutive or augmentative form, ex. *repetina* is both '*veliki rep*' (en. big tale) and '*velika repa*' (en. big beet). This direct definition of derivations inside the dictionary would definitely result in 100% precision, and a very high recall, but it may not detect any new entries in the text that language, as a living thing, makes possible. This is the main reason that we have opted to pursue another path i.e. building the morphological grammar for detection of diminutive and augmentative nouns in Croatian. The other reason falls in the domain of too many person hours needed to add each derivation manually directly to the dictionary.

4 Corpus

We have divided our corpus into 2 main categories in the following manner: CAT1 consists of children stories, and CAT2 of novels, mainly by Croatian authors. Bosanac et al. [4] have found six basic semantic and pragmatic features of diminutives in Croatian: the object in question is in diminutive form because it is considered small, seen with affection, considered negative (pejorative meaning), large or neutral. Additionally, the meaning can be contextualized or lexicalized. "Small" and "affectionate" were most prominent semantic connotations of diminutives. According to [7], diminutives are used as a politeness and softening device in discourse, so we can hypothesize that they would be more common in texts containing dialogues, such as novels.

Considering this, we have chosen two stylistically different corpora in order to analyze differences in the usage of stylistically specific words, i.e. diminutives and augmentatives, in different environments. We assumed that novels and stories directed towards children contain a larger amount of diminutives and augmentatives then texts directed towards adult audience, unless specifically needed in a context. Children corpus would most likely contain more words with "small" and "affectionate" meaning while novels may contain a number of diminutives and augmentatives for their poetic and stylistic meanings.

Table 1 represents the size of each of the two corpora. We have assumed that children's novels (CAT1) would contain more diminutives then regular novels (CAT2). This has been proven true, as we have found 7 times more diminutives by their relative frequency in CAT1 (0.96%) then in CAT2 (0.13%). However, although we expected to

---

[9] Some derivations (like *kamenina*) are only rarely used and found mainly in literal texts.

[10] Some of the examples are, however, limited to individual usage. Thus the example *kamenina* which Babić uses has been used by the literary writer Božić [2], while many average Croatian speakers would find it unusual and eccentric.

find significantly more augmentatives in CAT1 as well, their number was about the same in both corpora.

Table 1. The size and frequency of diminutives and augmentatives in the corpora constructed for testing purposes

Category	Number of tokens	Number of diminutives		Number of augmentatives	
CAT1	240 616	2 312	0.96%	114	0.047%
CAT2	504 876	696	0.14%	59	0.032%

For the purpose of this research, we have chosen to compare the results obtained manually (human annotators) and automatically (NooJ morphological grammar) in these two categories (CAT1 and CAT2). We present our findings in the Results section.

5   Recognizing Diminutive Nouns

The morphological grammar for recognizing diminutive Croatian nouns consists of four main graphs depending on the gender related endings. The first graph recognizes diminutives characteristic for the feminine nouns (including diminutive suffixes: *-ca, -ica, -čica*), the second graph for masculine nouns (*-ac, -ak, -čić, -ečak, -ičak*) (Fig. 1) and the third graph for neutral nouns (*-ce, -ance, -ašce, -ence, -ešce*).The fourth graph is for recognizing diminutives build with endings that are not related to gender and as such includes masculine, feminine and neutral nouns (*-ić, -če, -eljak, -erak, -uljak*). Still, the situation is not as clear as it may appear since there are some suffixes that are used in different gender nouns [1]. This is not only a characteristic for Croatian diminutives but it is also observed in other languages like Romanian [7], Slovene [17] or Slovak and Czech [15]. Babić [1] gives a list of 25 diminutive suffixes in total, three of which are highly productive (*-ica, -ić, -čić*), one is partially productive (*-če*) while others are considered to be weakly or nonproductive suffixes.

The main logic behind each of the graphs is to take the unknown word, divide it into three sections: N = main noun, S = diminutive particle, and case ending; and check the following:

1. does the section N exists in the main dictionary as the common noun in singular Nominative (in the gender defined by the subgraph type: feminine, masculine, neutral or any gender);
2. what is the diminutive particle S;
3. what is the case ending.

Finally, each recognized word is annotated as a common noun N, of type = *umanj* (short for diminutive in Croatian), and the case and number depending on the case ending defined in *c*, with the main noun N as its superlemma. Connecting diminutives and augmentatives with their superlemma will allow us to find all the forms of a noun *kuća* (en. house) whether it is written in the regular form or as a diminutive and augmentative form.

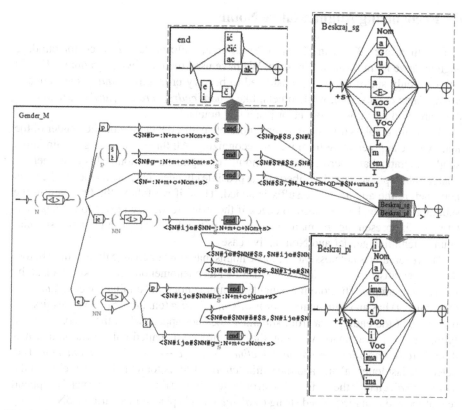

Fig. 1. Section of a morphological grammar recognizing diminutives characteristic of masculine nouns

The graph depicting the recognition of diminutives for masculine nouns (Fig. 1) is given here as an example since the same logic is followed in the remaining graphs for diminutives. This graph consist of more than one path (as one might expect). This is due to the fact that Croatian morphology requires some phonological alternations when particular phonemes are in the same environment, such as changing 'b' for 'p' (upper path in Fig. 1) or shortening the 'ije' to 'je' (lower paths in Fig. 1). So for example, diminutive for the noun 'cvijet' (en. flower) is 'cvjetić' (en. little flower).

Such alternations required separate approach in the grammar which resulted with some extra paths in the graph. Due to these and similar alternations, the section of a word marked as N is not always a root noun in its full form. Thus, the missing, or alternated phonemes had to be glued back to the N before we can check if the root noun exists in the dictionary. This has been done with a NooJ special character '#' that lets us concatenate more strings into one. Thus, the expression '$N#ije' adds the string 'ije' directly to whatever string is found in the variable $N. As it will be seen in the following section, we have utilized this feature in the grammar for recognizing augmentative nouns as well, since we were dealing with the same situations in detecting the root noun.

6   Recognizing Augmentative Nouns

In [16] there are 23 and in [2] even 30 different suffixes that are used for building augmentative noun forms in Croatian. We have used them all in our grammar (Fig. 2) although only six of these are considered to be very productive: *-ina, -čina, -etina, -urina, -jurina* and *-usina*, while others, like *-enda, endra, -erda, -(j)urda, -urenda* or *-čuga* are used for either only one or just a few nouns.

The main graph is divided into three possible paths, depending on the gender of the noun whose augmentative form we are recognizing. All three paths merge again into a single subgraph that describes the case endings of augmentative nouns. Inside the gender related subgraph we have described all the possible changes that may happen to the main noun before the augmentative suffix is added. Thus, if we follow the first path in the subgraph **Gender_F_**, we need to check if the string recognized prior to the augmentative suffix, presenting the main noun, exists in the dictionary as a common singular noun in feminine gender and Nominative case.

However, this is not possible for those nouns that loose or change their ending before the augmentative suffix is added. To illustrate this phenomenon, let us take a look at the following example. The common singular feminine noun in Nominative case *kabanica* (en. raincoat) has an augmentative form *kabani + četina* (en. large raincoat). At first we observe that the ending *–ca* from *kabanica* does not appear when the suffix is added. From Croatian Grammars we know that this process is called palatalization and it is described as: *kabanic-a → kabanic + etina → [c + e → č + e] → kabaničetina* (i.e. delete the last letter 'a' and add the suffix '*etina*'; if 'c' is found before the 'e' from the suffix '*etina*', change the 'c' into 'č'). In order to simulate this process, we need to append 'ca' at the end of a recognized string ('*kabani*') that is placed in a variable $N but only where 'č + etina' follows the variable $N. Again, by using NooJ notation described in the previous section **<$N#ca =: N+c+f+Nom+s>** we are able to check if a feminine common noun '*kabanica*' exists in our dictionary. Only if a noun from the variable $N#ending exists in our dictionary, we are able to proceed along our path which takes us to the content of variable $S. This variable holds an entire new subgraph *aug_F* that holds all the possible endings used for building augmentatives of feminine nouns.

After any of these suffixes is recognized in the current string, the word is annotated as an augmentative noun with the main noun as its superlemma. The case and number information are added in the following step after the string passes through the subgraph Medo_sg or Medo_pl where possible case and number endings are given (Fig. 2).

Our grammar is not retained from the ambiguity since different nouns may produce the same augmentative (and diminutive) forms. For example, augmentative ***kosturina*** may be recognized from the noun *kost* (en. bone) or *kostur* (en. skeleton) and augmentative ***maščurina*** may be recognized from the noun *mast* (en. grease) or *maska* (en. mask). Of course, the same would happen even if we added derivational paradigms directly to our dictionary entries. But, since lexical ambiguity is not something that we can deal with on this (morphological) level of analysis we do not consider this to be a downside of our grammar.

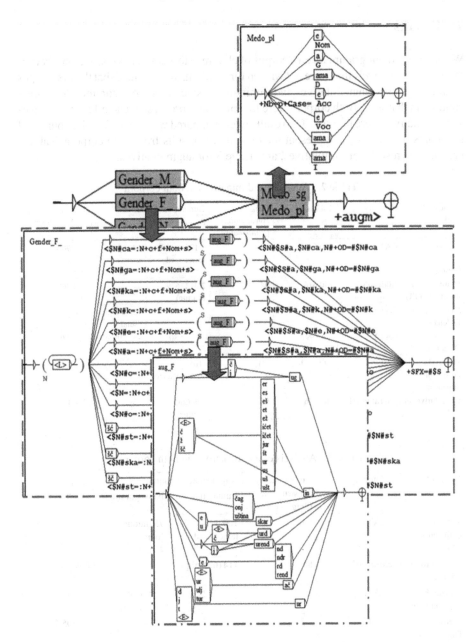

Fig. 2. Section of a morphological grammar recognizing augmentatives

7 Results

We have tested our grammars on a corpus of children stories and novels, and on a corpus that consists of novels that were not written for children. We assumed that the first corpus would have more augmentatives and diminutives, since it is more common to address children in that way. Both corpora were first manually processed, and all augmentatives and diminutives were marked. The results were compared with those found by our NooJ grammars. Firstly, we will present the quantitative results from both corpora. Table 2 presents the results for diminutives, and Table 3 for augmentatives.

Table 2. Analysis of diminutives in both corpora

Name of the work	Category	Tokens	Dim. constructions (*"little house"*)	Diminutives found		% of Diminutives
				Cumulative	NooJ grammar	
Various children's stories	CAT1	240 616	94	2 195 + 117 lexicalized	549 unique found, 538 correct	0.96%
Cumulative for children's novels (CAT1)		**240 616**	**94**	**2 312 (679 unique)**		**0.96%**
J. Polić Kamov: Isušena kaljuža	CAT2	104 585	52	147	255 unique found, 241 correct	0.14%
J. Kozarac: Đuka Begović	CAT2	33 992	3	80		0.23%
V. Novak: Posljednji Stipančići	CAT2	366 299	21	105 + 59 lexicalized		0.044%
Cumulative for general novels (CAT2)		**504 876**	**55**	**696 (283 unique)**		**0.14%**

Table 3. Analysis of augmentatives in both corpora

Name of the work	Category	Augmentative constructions (*"big house"*)	Augmentatives found		% of augmentatives
			Cumulative	NooJ grammar	
Various children's stories	CAT1	60	100 + 14 lexicalized	75 unique found, 41 correct	0.047%
Cumulative for children's novels		**60**	**114 (63 unique)**		**0.047%**
J. Polić Kamov: Isušena kaljuža	CAT2	13	52	56 found, 45 correct	0.049%
J. Kozarac: Đuka Begović	CAT2	3	7		0.0205%
V. Novak: Posljednji Stipančići	CAT2	54	15 + 87 lexicalized		0.0278%
Cumulative for general novels		**16**	**59 (51 unique)**		**0.032%**

We have separately counted augmentative and diminutive constructions like *velika kuća* (en. big house) and *mala kuća* (en. little house) which are not morphologically diminutives nor augmentatives, but semantically would fit into these categories. Our NooJ grammars did not detect them, as they are not considered augmentatives nor diminutives, but they are noted in the table (Tables 2 and 3) for reference.

After comparing human annotations with those yielded by our grammars, we have found the following results. Table 4 represents precision, recall and F-measure for children novels corpus (CAT1), while Table 5 represents the same statistical measures for general novels corpus (CAT2).

Table 4. The results for children novels corpus

	Precision	Recall	F-measure
Diminutives	0.9800	0.8239	0.8952
Augmentatives	0.5467	0.6508	0.5942
Overall	**0.7633**	**0.7373**	**0.7447**

Table 5. The results for general novels corpus

	Precision	Recall	F-measure
Diminutives	0.9451	0.8516	0.8959
Augmentatives	0.8036	0.8824	0.8411
Overall	**0.8743**	**0.8670**	**0.8685**

From the data in Tables 4 and 5 we can conclude that our grammar for detecting diminutives [f-measure: 0.90] performs better than the grammar for detecting augmentatives [f-measure: 0.72]. And if we are to consider both grammars to function as one system, its scores would be over 0.80 [P: 0.82; R: 0.80; f-measure: 0.81], which we found satisfactory for this first try to dealing with such derivational forms.

So, what have we learned from our data and how can we implement that knowledge? Although our preliminary results are in general satisfactory, an analysis of an error typology is necessary in order to improve the grammar, and consequently, its results. After a thorough evaluation, we can offer the following typology of errors:

1. personal names: first names (*Slavica, Ančica*) and last names (*Ilačić, Krpina*)
2. Toponyms: *Kučine, Harmica*
3. Possessive adjectives: *Katino* (en. Kate's), *maćehino* (en. stempother's)
4. Word forms: *sokak* (**not** a small juice (*sok*=juice)), vrtuljak (**not** a small garden (*vrt*=garden))
5. Typos: *piace* (**wrong:** pi+ac+e), *proizvođačaka* (**wrong:** proizvođač+ak+a)

Personal names can be, as shown in Introduction, be diminutive in their origin (*Slava – Slavica, Ana – Ančica*). However Croatian speakers often do not treat them as diminutive forms, but as regular noun forms, similar to example (*ploča,* panel – *pločica* – bathroom tile). Next to that, not all personal names are diminutives or augmentatives by formation, particularly last names. The same can be said about toponyms. Possessive adjectives formed from feminine nouns are another category that came back as augmentative forms in the first results, although they are clearly not.

In the fourth type of errors, although both words are correctly segmented (nouns '*sok*' and '*vrt*' exist in Croatian dictionary and all the suffixes and case endings are valid), their diminutive forms are not '*sokak*' and '*vrtuljak*' but rather '*sokić*' and '*vrtuljčić*' respectfully. In addition, words '*sokak*' and '*vrtuljak*' also exist in Croatian language but their meanings are '*street*' and '*merry-go-round*'. The reason why these words were recognized by our grammar is that they were not in our NooJ dictionary.

Errors of the fifth type are similar to the previous type of errors in that the words are segmented correctly (there really are words '*pi*' and '*proizvođač*' in the dictionary, and both 'ac' and 'ak' are regular suffixes for diminutives, and 'e' and 'a' are case endings that may appear after these suffixes). Still, these words do not exist in Croatian language and as such are marked incorrect.

The grammar for recognition of diminutives and augmentatives is marked as a grammar with low priority level, which means that if the word is recognized by the dictionary, the grammar will not be applied to that word. Thus, errors of type 1, 2 and 4 are easily solved by adding the missing names, toponyms and other word forms. Similarly, if a grammar for recognizing possessive adjectives is build and applied to the text prior to the grammar for recognizing diminutives and augmentatives, it would solve the errors of type 3. For the last type of errors, however, we do not offer any solution that would be possible on this level of text analysis.

8   Conclusion

In this paper we have presented the NooJ grammar-based solution for recognition of diminutives and augmentatives in the Croatian language. We started this project with a thorough review of the literature in order to analyze the morphology of Croatian diminutives and augmentatives. In the Croatian language, diminutives and augmentatives are formed from the base word with a number of suffixes, such as -ić and-ica for diminutives or -ina for augmentatives. Based upon this data, our grammars were constructed. Their task was to recognize and mark any diminutive or augmentative form, in case that the base word was already in the dictionary (e.g. *stol*, table -> *stolić*, little table). We have tested the grammars on two stylistically different corpora, the one based on children novels, and the one based on general novels in Croatian language.

Except for finding data that helped us evaluate our grammars, we have also collected data about the frequency of diminutives and augmentatives in the Croatian language. In the case of diminutives, we have found them to be most common in children novels, comprising almost 1% of all words in the text, while they accounted for only 0.14% of tokens in general novels corpus. As for the augmentatives, they are a very rare word form in texts, comprising less than 0.05% of tokens in both corpora.

Our grammars have successfully found a large percentage of diminutives, with average precision over 96%, recall over 83% and F-measure of 89%. In the case of augmentatives, the numbers were somewhat lower, due to the large number of homographic words. In the Croatian language, a word can be e.g. a possessive adjective with the suffix -*ina*, which is the most common augmentative suffix. Due to the large number of false positives, the grammar for recognizing augmentative forms yielded the average precision of 67%, recall of 76.6% and F-measure of 71.7%. Although considerably lower than in the case of diminutives, it should be noted that these results are based on a much smaller number of words, as augmentatives are much rarer in Croatian texts.

It is a matter for discussion if this manner of processing diminutives and augmentatives in Croatian is indeed the best method, as we have to take into account many homographic words which need to be differentiated from true results, the non-standard forms, and those based on words which are not already in the NooJ dictionary. We can propose that the best method for a complete recognition system would be both the construction of NooJ grammars, and manual addition of derivational paradigms directly to the NooJ dictionary. However, as we have seen that the relative frequency of diminutives, and especially augmentatives, is considerably low, this manual method may be too expensive and too slow. At this point, the grammars recognize almost all the diminutives and majority of the augmentatives. Thus, it is our strong belief that this work can be of referential benefit to future researchers of Croatian diminutives and augmentatives.

References

1. Babić, S.: Sustav u tvorbi hrvatskih umanjenica, (System for building Croatian Diminutives). In: Slavistična revija, Letnik 20:1, Ljubljana (1972). (in Croatian)
2. Babić, S.: Sufiksalna tvorba uvećanica u hrvatskome književnome jeziku, (Suffix Word Formation of Augmentatives in Croatian Literary Language). In: Suvremena lingvistika, 41–42(1–2), pp. 11–20 (1996) (in Croatian). http://hrcak.srce.hr/23938
3. Babić, S.: Tvorba riječi u hrvatskom književnom jeziku. (Word Formation in Croatian Literary Language). Zagreb: Hrvatska akademija znanosti i umjetnosti: Nakladni zavod Globus, pp. 215–221 (2002). (in Croatian)
4. Bosanac, S., Lukin, D., Mikolić, P.: A Cognitive Approach to the Study of Diminutives: The Semantic Background of Croatian Diminutives. Rector's Award paper, Zagreb: Filozofski fakultet, academic year 2008/2009 (2009)
5. Grandi. N.: Development and spread of augmentative suffixes in the mediterranean area. In: Ramat, P., Stolz, T. (eds.) Mediterranean Languages. Dr. Brockmeyer University Press, Bochum, pp. 171–190 (2002)
6. Grandi, N.: Renewal and innovation in the emergence of indo-european evaluative morphology. In: Körtvélyessy, L., Stekauer, P. (eds.) Diminutives and Augmentatives in the Languages of the World. Lexis: e-J. Engl. Lexicology 6, 5–25 (2011). http://lexis.revues.orgimg/pdf/Lexis_6.pdf
7. Hornoiu, D.: Avoiding disagreement in Romanian conversational discourse: the use of diminutives. In: Bucuresti, A. (ed.) Univerității Bucuresti, Editura, pp. 99–105 (2008)
8. Jovanović. V.: Deminutivne i augmentativne imenice u srpskom jeziku. Beograd: Institut za srpski jezik SANU (2010). (in Serbian)
9. Katunar, D.: Diminutives in Action: A cognitive account of diminutive verbs and their suffixes in Croatian. Suvremena lingvistika 39(75), 1–23 (2013). http://hrcak.srce.hr/105494
10. Körtvélyessy L.: A cross-linguistic research into phonetic iconicity. In: Diminutives and Augmentatives in the Languages of the World, Lyon, pp. 27–40 (2011). http://screcherche.univ-lyon3.fr/lexis/IMG/pdf/Lexis_6.pdf

11. Krstev, C., Vitas, D.: Extending the Serbian E-dictionary by using lexical transducers. In: Koeva, S., Maurel, D., Silberztein, M. (eds.) Formaliser les langues avec l'ordinateur: De INTEX à NooJ. Cahiers de la MSH Ledoux. Presses Universitaires de Franche-Comté, France, pp. 147–170 (2007)
12. Marković, I.: Uvod u jezičnu morfologiju. Disput, Zagreb (2013). (in Croatian)
13. Mitrićević-Štěpánek, K.: Deminutivi u funkciji nabrajanja i konfrontacije u českom i srpskom jeziku. In: Opera Slavica XVII, 4 (2007). (in Serbian). https://digilib.phil.muni.cz/bitstream/handle/11222.digilib/117151/2_OperaSlavica_17-2007-4_4.pdf?sequence=1
14. Mota C.: Portuguese morphology with INTEX 4.33. In: Koeva, S., Maurel, D., Silberztein, M.: Formaliser les langues avec l'ordinateur : De INTEX à NooJ. Cahiers de la MSH Ledoux. Presses Universitaires de Franche-Comté, France, pp. 135–146 (2007)
15. Panocová, R.: Evaluative suffixes in slavic. Bull. Transilvania Univ. Brasov, Series IV: Philology Cult. Stud. 4 (53), No. 1, 175–182 (2011)
16. Pintarić, N.: Kontrastivno rječotvorje: imenička tvorba u tablicama, Zagreb: Filozofski fakultet, Odsjek za zapadnoslavenske jezike i književnosti (2010). (in Croatian)
17. Sicherl, E., Žele, A.: Nominal diminutives in Slovene and English. In: Linguistica (Ljubljana), številka letn. 51, pp. 135–142 (2011)
18. Silberztein, M.: NooJ Manual 2003. www.nooj4nlp.net
19. Spasovski, L.: Morphology and Pragmatics of the Diminutive: Evidence from Macedonian. Doctoral Thesis, Arizona State University (2012)
20. Vučković, K., Tadić, M., Bekavac, B.: Croatian language resources for NooJ. J. Comput. Inf. Technol. **18**, 295–301 (2010). CIT

Inflectional and Morphological Variation of Arabic Multi-word Expressions

Dhekra Najar[✉], Slim Mesfar, and Henda Ben Ghezala

RIADI, University of Manouba, Manouba, Tunisia
dhekra.najar@gmail.com, mesfarslim@yahoo.fr, hhbg.hhbg@gmail.com

Abstract. CompounDic is an Arabic MWEs dictionary that lists many entries, divided into more than 20 domains. It lists only MWEs in their base form. With regard to syntactic and morphological flexibility, the lexicon covers 2 types of MWEs: Fixed MWEs (no variation allowed) and semi-fixed MWEs (variation in their structural pattern). Arabic presents distinctive features to deal with MWEs processing. A lot of possible derivations are possible (plural or dual forms, multiple irregular plurals). In addition, we need to process agglutination forms. In this paper, we will study the structural variability of semi-fixed multiword expressions in Arabic language in order to recognize the morphological and inflectional variations. We will adopt a recognition approach based on the use of a cascade of local grammars.

The recognition system is based on NooJ's local grammars as well as an Arabic MWEs dictionary covering more than 20 domains. The inflectional and derivational rules, which concern semi-fixed MWEs, use some specific morphological operators that will be described as well. Finally, we present new results showing the experimentation scores of morpho-lexical coverage enhancement.

Keywords: Multi-word expressions · Natural language processing · NooJ · Arabic language · Compound words variation

1 Introduction

A Multi-Word Expressions (MWEs) are groups that work together as units to express a specific meaning. They can be formed by combining two or more words together. Generally, lexical and morphological analyzers are not able to recognize multiword expressions unless they are listed in internal resources. Automatic analyzers usually process MWE as separated terms. As a result, semantics is lost because generally the meaning of the MWE is different from the meanings of its components.

Most multi-word expressions allow certain types of variability on their components. This problem has to be taken into account for their description to be able to recognize them in texts as well as their potential variations.

The identification of MWEs is essential for any natural language processing based on lexical information. Therefore, recognizing only the limited MWEs that are usually listed in computational lexicon is not enough. The morphological and inflectional variability of MWEs and their lexical particularities need to be described in the computational lexicon

© Springer International Publishing AG 2016
L. Barone et al. (Eds.): NooJ 2016, CCIS 667, pp. 37–47, 2016.
DOI: 10.1007/978-3-319-55002-2_4

in order to be able to recognize the full range of their occurrences in texts. The rest of the paper is organized as follows: Sect. 2 describes expressions topology as well as their structural variability and presents the MWE"s lexicon CompounDic. The proposed approach is discussed in Sect. 3. Section 4 shows the experimental results. Section 5 summarizes the results of this work and draws conclusions.

2   Multi-word Expressions

2.1   Arabic MWE's Variability Types

Based on previous works, we identify three types of variability of MWEs: fixed, semi-fixed and syntactically flexible.

- Fixed MWEs are considered as a list of words with spaces and with no morphological variation allowed. This category contains unambiguous compound expressions such us (Middle East, الشرق الأوسط) and frozen sentences such us pragmatically fixed expressions (مدى الحياة, forever) and proverbs.
- Semi-fixed expressions allow variations including graphical variants, which are the graphic alternations between the letters (ي, ى) and the letters (ه, ة), as the following illustrates. As well, many morphological variants can effect semi-fixed expressions. Specifically, we mention variations that express person, number, tense, gender, and the definite article that is carried out by the fixed morpheme (ال, Al) (Fig. 1).

Fig. 1.   Example of inflectional variants of an entry.

- While MWEs that are syntactically flexible allow new external elements (components) to intervene between the MWE components (Fig. 2).

Arabic words are characterized by their complex structure. In comparison with Semitic languages, Arabic language presents distinctive features, namely the vocalization that causes a lexical ambiguity in texts. Also, Arabic is an agglutinative language (the prefix (definite article (the, ال), prepositions (for, ل) and (with, ب), conjunctions (and, و), suffixes (her, ها)).

The Arabic language has a complex MWEs structure (up to 5 units) and a lot of possible variations and derivations (dual forms, multiple irregular plurals... ect). The recognition of all potential inflected and agglutinated forms attached to each entry needs

Fig. 2. Example of syntactically flexible of an entry.

a special tokenization that depends on their linguistic specificities. However, we used to make some specific tools to be able to deal with the specificities of the Arabic language.

Arabic presents distinctive features to deal with MWEs processing. A lot of particular variations are possible:

- Agglutinated forms;
- Inflectional variations: (Gender and number: plural or dual forms, multiple irregular plurals).
- Morphological Variations: (Definite article, Personal agglutinated pronouns, Agglutinated conjunctions and prepositions).

2.2   CompounDic

In previous work, we have semi-automatically built CompounDic (Najar et al. 2015), an Arabic 2 units MWEs thematic lexicon. For this purpose, we have taken advantage of NooJ's[1] linguistic engine strength in order to create this large coverage terminological MWEs dictionary for Modern Standard Arabic language CompounDic. NooJ is a linguistic development environment that allows formalizing complex linguistic phenomena such as compound words generation, processing as well as analysis.

However in Nooj "simple words and multi-words units are processed in a unified way: they are stored in the same dictionaries, their inflectional and derivational morphology is formalized with the same tools and their annotations are undistinguishable from those of simple words" (Silberztein 2005).

CompounDic contains 36960 entries classified into more than 20 semantic domains. It covers the category of fixed expressions except proverbs and semi-fixed expressions as well as the different types of MWEs such as expressions that are traditionally classified as idioms, prepositional verbs, collocations, and so on. In this lexicon, we didn't deal with flexible expressions.

All the entries of CompounDic are manually set in the base form: "indefinite singular form". Then, all the listed MWEs are voweled manually so that NooJ would be able to recognize unvoweled, semi-voweled as well as fully voweled MWEs. The manual vocalization is an extremely important step since it allows to vowel entries depending

---

[1] http://www.nooj4nlp.net/.

on their semantic information since we can find a word that has different way of vocalization and different meanings. This helps reducing linguistic ambiguities in Arabic texts.

The final manual step is classifying the MWEs according to 2 criteria: the grammatical composition (N1 N2), (N1 ADJ, and so on).

In fact, the Arabic MWE can be a combination of different forms: a verb, a noun, an adjective and a particle. Most of MWEs are composed of one or more nouns (N), adjectives (ADJ), adverbs (ADV) or simple named entities. We provide the syntactic phrase structure composition of our Arabic MWEs, giving each entry of our lexical resource its component elements (noun + noun, noun + adjective, verb + preposition + noun...).

We manually extract a list of about 15 patterns of MWEs compositions classified into 4 basic categories (Table 1):

Table 1. Patterns of MWEs compositions

Type	Structure
Descriptive compound	ADJ_N
	ADJ_prep
	ADJ_prep_N
	ADJ1_ADJ2
Compound nouns	N1_N2
	N1_prepN2
	N_ADJ
	N_prep
Negation	Neg_N
	neg_V
Prepositional nouns	prep_N
Compound Verbs	V_N
	V_prep
	V_prepN
	V1_V2

The entries of CompounDic are classified into more than 20 domains as shown in Table 2.

Every entry in CompounDic is stored with information about its structure, number of units and domain. To give a simple example from the technical domain in our lexicon:

وإلتهام اللّزّان **N + Structure = N1_N2 + CMPD + Units = 2 + Domain = Technical.**

As it was said, fixed MWEs always occur in exactly the same structure and can be easily recognized by a lexicon. However, most MWEs allow different types of

Table 2. Number of entries in CompounDic per domain

Domain	Entries	Domain	Entries
sportive	1434	economical	2222
financial	2268	media	508
agriculture	2401	educational	1526
political	3251	religious	1645
press	44	UN	1582
military	2323	Touristic	2251
legal	2191	Computer	2058
psycological	2068	weather	329
social	2572	transport	2356
Industry	582	engineering	1277
administrative	578	technical	2076
Total			36960

modifications. In Arabic language, we can reach an average of 33 possible variations to each MWE entry. Arabic presents distinctive features to deal with MWEs processing such as plural or dual forms, multiple irregular plurals and agglutination forms. With this in mind, we still have a lot of possible variations to recognize from CompounDic lexicon.

3 Approach

In order to improve Natural Language Processing system performances, it is important to identify MWEs in texts since it helps to disambiguate semantic and lexical content. Generally speaking, we have 2 potential solutions to recognize CompounDic entries variations:

- **Generation method:** focuses on inflectional and derivational descriptions that are manually implemented for each MWE entry. This method is not efficient due to the exponential complexity that can cause and the time that take to manually implement descriptions.
- **Recognition method:** focuses on lexical grammars that recognize the MWE's variations. This method uses local grammars to recognize the related forms of CompounDIC entries without generating them. Usually, the result of the recognition method is precise. Furthermore, it processes agglutinated forms. However, we will be faced to heavy linguistic analysis since NooJ will check the lexical constraints for each digram.

In view of this, we propose to use the recognition method with based-rules local grammars in order to automatically recognize the inflectional and morphological variations from CompounDic entries using NooJ's linguistic engine. We are going to add some enhancement to this method in order to avoid heavy linguistic analysis especially while processing big corpus.

To sum up, our system will be able to:

- Recognize the morphological and inflectional variations of Arabic MWEs.
- Annotate MWEs in text with their distributional (Domain = Financial…) and syntactic information (Noun + Noun, Noun + Adj…).
- Get a better semantic representation.
- Reduce the lexical and syntactic ambiguity.

4   Grammar

We are going to use NooJ's linguistic engine to implement a local grammar describing the structural variability of Arabic MWEs. This grammar will be able to recognize all the morphological and inflectional variants of CompounDic entries, namely:

- Gender (female, male);
- Number (dual, plural);
- Definite article: the fixed agglutinated morpheme (ال, Al);
- Personal agglutinated pronouns;
- Agglutinated conjunctions and prepositions (for, ل), (with, ب), (and, و).

As noted earlier, the enhancement of the recognition method is important to avoid heavy linguistic analysis. For this reason, we are going to focus the analysis on the units that are attested to be a part of a MWE.

- Step 1: extract all the units of our CompounDIC.
- Step 2: add to the extracted units in El_DicAr the distributional information (+CmpElem).

To do this, we have developed a program to enrich El-Dicar[2]. It allowed us to add semi automatically about 2000 unknowns (technical words) and automatically 7000 distributional information (+CmpElem). We are still working on the enrichment of El-DicAr dictionary.

We illustrate this semi-automatic enrichment program by Fig. 3.

---

[2] Electronic Dictionary for Arabic "El-DicAr" resources (Mesfar et al. 2008), developed using NooJ's linguistic engine.

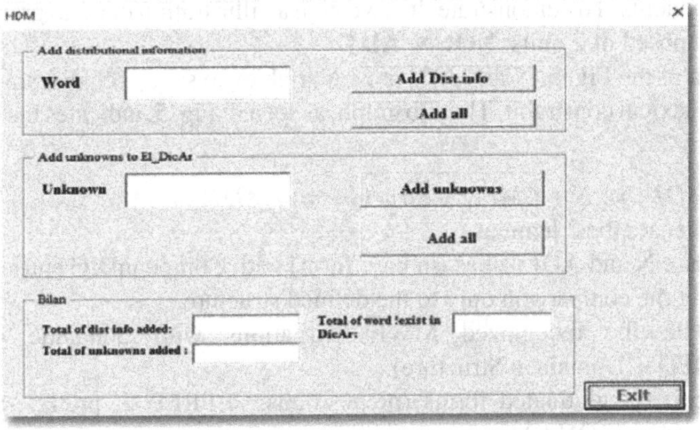

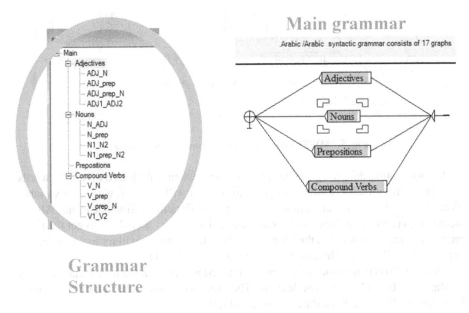

Fig. 3. Enrichment program platform.

Our local grammars are implemented based on the 17 patterns of MWEs compositions that we have extracted as shown previously. As we can see in Fig. 4 we have the grammar structure that shows all the MWEs structures and the main grammar of our system.

Fig. 4. Local grammar structures and the main sub graph.

With the distributional information +CmpElem, the linguistic analysis of our grammar will be limited on the units that are attested to be a part of a MWE. To do so, we are going to use distributional information +CmpElem in the grammar to identify

MWEs components. To demonstrate this, we give an illustration of a sub graph of MWE structure composed of 2 units: **NOUN_ADJ**.

As shown in the Fig. 5, N and ADJ are 2 Variables to save each digram element to use them in a lexical constraint. The sub graph, as seen in Fig. 5, indicates the constraints below:

1. $\$N\_\# \#\$ADJ\_ = : N + CMPD + Structure = N\_ADJ$[3]
   - Concatenate the 2 lemmas.
   - Compare N and ADJ values (in base form) with CompounDIC entries.
   - Restrict the comparison only to the defined structure.
   - Annotate the recognized MWEs variations with Semantic description (+CMPD + Domain + Structure).
   - Recognize agglutinated forms (prepositions: < PREP >, prefix: < PREF >, pronoun: < PRON >).
2. $\$N\_ \$ADJ\_, N\$1S>$
   - $\$N\_$: Represents the lemma of the lexical unit stored in $\$N$ variable
   - N$1S: inherits the semantic information (Domain) from the recognized MWE to annotate the matching sequence.

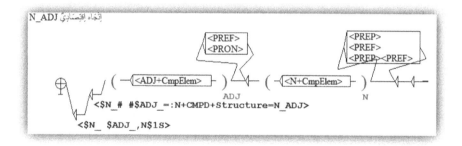

Fig. 5. MWEs variations local grammar NOUN_ADJ

Demonstrating this, the grammar process a text when it finds two or simple words with the distributional information (+CmpdElem): it will put each word in a variable $\$Var\_$ tracked by "_" to set them to their base form (indefinite Singular form). All the stored consecutive variables will be concatenated < $\$Var1\_ \$Var2\_$ > to get the same multi-word expression but in the base form. Then, the grammar will try to find a similar entry of the MWE in our lexicon using the first constraint (1).

Once the MWE is found, it will be recognized and considered as a variation of an existing MWE in CompounDic lexicon. The grammar allows inheriting the semantic information (Domain) from the recognized MWE.

However, we have a particular case of entries containing agglutinated prepositions (V_prepN, N1_prepN2, ADJ_prepN) as shown in the sub graph below. It's not possible for our grammar to recognize agglutinated MWE elements. So, we have made some changes in the constraints of sub graphs of MWEs structures with agglutinates elements.

---

[3] NooJ's syntactic, inflectional and semantic categories are detailed in Annex.

To be specific, we give the example of the prepositional structure NOUN1_prepNOUN2 (Fig. 6).

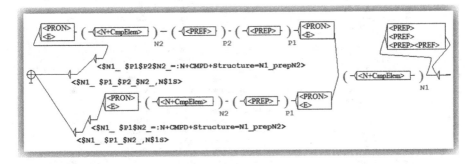

Fig. 6. MWEs variations local grammar NOUN1_prepNOUN2

The same thing with the first example except:

1. $N1_ $P1$P2$N2_ = : N + CMPD + Structure = N1_prepN2
   Concatenate the 2 lemmas including the prepositions (without modifying the form of the prepositions). Check in CompounDIC entries. If it exists in our lexicon then it will be considered as a variation of MWE.
2. *$N1_ $P1_ $P2_$N2_, N$1S*
   - N$1S: inherits the semantic information (Domain) from the recognized MWE to annotate the matching sequence.

5   Results and Discussion

To test the lexical recognition of our grammar, we launched the linguistic analysis of our test corpus. We presented preliminary experiments on a corpus containing 870 heterogeneous articles from Internet. We reported high quality result.

The table above presents the recall and precision obtained by testing the grammar on the test corpus. The results, as seen in Table 3, indicate that we have reached high quality results of recognition. Our results in term of precision (0.97 of precision) are better than other existing approaches. We presented preliminary experiments on a Concordance:

Table 3. Results

Precision	Recall
0.97	0.88

We believe that this automatic method ameliorated the precision of the results by recognizing all MWEs forms in the text (Fig. 7).

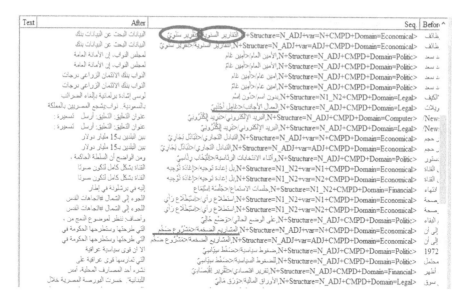

Fig. 7. Concordance

Illustrating the concordance, our grammar recognized expressions such as:

- (المشاريع الضخمة, **huge projects**): definite expression in the plural.

The base form of this expression in our lexicon is (مشروع ضخم, huge project).
Several obstacles make the recognition of Arabic MWE's variations really complicated such as high inflectional nature, morphological ambiguity related to some agglutinated forms, variant sources of ambiguity (unvoweled texts...) and dual forms for
pronouns and verbs. These specificities of Arabic language represent the most challenging problems for Arabic NLP researchers.

More specifically, the silence in our grammar is due to some problems in
CompounDic lexicon such as:

- False vocalization of words such as (misplaced vowels);
- Common typographical errors such as confusion between Alif and Hamza or the
  substitution of (ه, ة) and (ي, ى) at the end of the word;
- Lexical ambiguity of some agglutinated forms;
- Lack of entries in our lexicon.

6   Conclusion

MWEs are combinations of single terms expressing various meaning compared to the
combination of single word's meanings. This paper focuses on recognizing multi-word
expressions inflectional and morphological variations in Arabic corpus. Our research

has shown that rule-based approaches are more efficient in recognizing the entire multi-word expressions variations, especially morphological variations. We believe that this automatic method has improved the precision of the results.

Further research is needed to better understand the topology of MWEs in different languages.

7 Annex

NooJ's syntactic categories:

Syntactic codes	
<ADJ>	Adjective
<V>	Verb
<N>	Noun
<ADV>	Adverb
<CONJ>	Conjunction
<PREP>	Preposition
<PREF>	Prefix
<PRON>	Pronoun
<REL>	Relative pronoun
<PART>	Particle
<E>	Empty caracter
<P>	Ponctuation
Inflectional codes	
<s>	Singular
<p>	Plurial
<m>	Male
<f>	Female
Semantic codes	
<CmpdElem>	Component of a MWE

References

Najar, D., Mesfar, S., Ghezela, H.B.: A large terminological dictionary of Arabic compound words. In: Okrut, T., Hetsevich, Y., Silberztein, M., Stanislavenka, H. (eds.) Automatic Processing of Natural-Language Electronic Texts with NooJ, pp. 16–28. Springer, Cham (2015)

Mesfar, S.: Analyse Morpho-syntaxique Automatique et Reconnaissance Des Entités Nommées En Arabe Standard. Thesis, Graduate School - Languages, Space, Time, Societies. Paris, France (2008)

Silberztein, M.: Nooj's dictionaries. In: Vetulani, Z. (ed.): Proceedings of the 2nd Language and Technology Conference. Wydawnictvo Poznańskie Sp. z o.o., Poznan (2005)

Quechua Module for NooJ Multilingual Linguistic Resources for MT

Maximiliano Duran$^{(\boxtimes)}$

Université de Franche-Comté, Besançon, France
duran_maximiliano@yahoo.fr

Abstract. The aim of this paper is to present the content of the first version of a Quechua linguistic module for NooJ and to describe its main components. They include several electronic dictionaries of verbs, nouns, and other POSs. The article also describes an ongoing work in the elaboration of a large set of paradigms which formalize Quechua inflectional and derivational morphology as well as morphological and syntactic grammars. We present the morphological codes used in the dictionaries, some graphs representing the numerals, the derivation and finally a demo text.

Keywords: Quechua · Quechua for NooJ · Quechua linguistic module · Quechua-French bilingual resources · Morphological grammars · Inflectional and derivational paradigms · Electronic dictionaries · Quechua simple verbs

1 Introduction

The linguistic resources presented here are intended to support the description of phenomena occurring in Quechua-French and Quechua-Spanish projects.

NooJ4Qu is a set of linguistic resources developed with NooJ for the automatic processing of Quechua. We follow the structure of the NooJ modules developed in the last ten years for various languages (Barreiro 2008 for Portuguese; Bogacki 2008 for Polish; Chadjipapa et al. 2008 and Gavrillidou et al. 2008 for Greek; Georganta and Papadopoulo 2012 for Ancient Greek; Aoughlis et al. 2014 for Tamazight; Dobrovolic 2014 for Slovene). It is composed of a demo text and contains two types of elements necessary for the linguistic analysis of texts. The first one comprises some dictionaries of lemmas which can operate independently. The second one is a system containing an important number of local grammars describing inflections and derivations obtained using nominal and verbal suffixes. It also includes a local grammar describing numerals ranging from 1 to 999999, a grammar that recognizes and annotates the dates and a grammar for the recognition and annotation of future forms.

It is not the purpose of this module to give users exhaustive dictionaries and complete libraries of local grammars but rather to show how Quechua, with its very rich morphology and specific peculiarities fits in NooJ's linguistic development environment.

L. Barone et al. (Eds.): NooJ 2016, CCIS 667, pp. 48–63, 2016.
DOI: 10.1007/978-3-319-55002-2_5

2 The Main Components of NooJ4QU

The lexical analysis of a text makes use of the following components launched simultaneously:

- Dictionaries of nouns, adjectives and non-verbal POSs,
- A dictionary of Quechua verbs,
- A library of morphosyntactic grammars and of inflectional and derivational rules.

    The files describing the inflection of the inflected entries are:

- Ngrammars.flx corresponding to the compiled dictionary Nquechua.nod,
- Vgrammars.flx corresponding to Vquechua.nod,
- A Quechua text containing 8 tales.

3 The Noun Electronic Dictionary

The electronic dictionary of simple nouns contains 1,472 entries. It does not include proper nouns. All of the entries have an inflectional paradigm assigned to them, represented by FLX=. For instance the entry unit *wasi* (house) inflects according to the paradigm class NVOCAL, thus the entry appears: *wasi*,N+FR="maison"+FLX= NVOCAL. Inflectional paradigms are standard pattern models or prototypes based on morphological suffixation rules. These rules cover variation in number, diminutives and superlatives, verbalizations, cases.

Quechua is a highly inflectional language:

- *llaqta-cha*   « small town »
- *kusillu-hina*   « resembles to a monkey »
- *Paris-ña*   « it's Paris already »
- *rima-sqa-y-man-ta*   « in what it concerns my speech »

The category of case is represented by the following set of suffixes {*-hina, -kama, -man, -manta, -nta, -ninta, - ninka, -nka, -p, -pa, -pi, -paq, -pura, -rayku, -ta, -wan, -y!, -niy!*}. The verbalization of a noun is obtained by one or more of the following derivation suffixes: {*-y, -yay, -chay*}

3.1 Nominal Inflections

There is another subset which is classified as enclitics: {*-ña, -raq, -puni,* }

The thousands of nominal forms that can be obtained by the inflectional paradigms are generated by the following set of nominal suffixes and their combinations in different layers:

SUF_N = {-ch , -chá, -cha ,-chik, -chiki, -chu, -chu?, -hina, -kama, -kuna, -lla, -má, -man, -manta, -..., -pa, -paq, -pas, -pi, -poss(7v+7c), -puni, -pura, -qa, -rayku, -raq , -ri, -s, -si, -sapa , -su, -ta, -taq, -wan, -y!, -ya!, -yá, -yupa , -yuq} (the

signs ?, ! are not part of the suffix, they are put only to indicate the presence of intonation).

The tag poss (7v+7c) represents the two sets of possessive suffixes:

- poss (7v)={-y, -yki, -n, -nchik, -yku, -ykichik, -nku} applied to nouns with a vowel ending.
- poss (7c)={-niy, -niyki, -nin, -ninchik, -niyku, -niykichik, -ninku} applied to nouns with a consonant ending.

They give us the following paradigms of possessive forms:

POSSV = i/POS+s+1 | iki/POS+s+2 | iku/POS+PEX+1 | nku/POS+p+3;
POSSAIC = nin/POSAI+s+3 | ninchik/POSAI+PIN+1| niikichik/POSAI+p+2;

The morphosyntactic grammars of dimension 1 (only a single suffix intervene) for the inflection of nouns is shown in Fig. 1.

Morpho-syntactic inflection grammar of dimension 1 for nouns

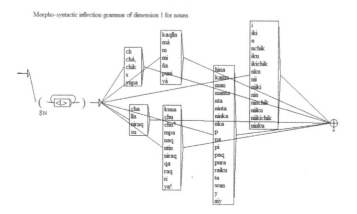

Fig. 1. Local grammar of one-dimension nominal inflection

For the combination of two nominal suffixes, we have developed an algorithm (See Duran 2013a which allows us how to obtain 517 valid agglutinations, some of which appear in the following paradigm:

N_V_2=:DCHACH I:DCHACHÁ I:DCHACHUI I:DCHACHU I:DCHACHIKI I:DCHAM
I:DCHAGEPA   ...   I :DCHALLA I:GEPACHUN I:GEPACHIK I:GEPACHU
I:GEPAGEPA I:GEPAKAMA I:GEPAKUNA I:GEPALLA I:GEPAMAA I:GEPAMAN ;

For the three-layer combinations of nominal suffixes we have obtained 2,108 valid
ones. And for the four-layer of four suffixes we obtain less than 1,800.

Nominal inflections like: *wasi-cha-lla-ikichik-paq-hina-raq*, "it is something that
would look nice in your house" are generated by paradigms containing six agglutinated
suffixes like:

CHALLA_1 = <E>/N+s I :CHALLAIKICHIKPAQHINARAQ;
CHALLAIKICHIKPAQHINARAQ= :CHA I:LLA I:IKICHIK I:PAQ I:HINA I:RAQ;

Grammatical gender is not proper to Quechua. Natural gender is differentiated in
lemmas, e.g. *maqta* boy, *pasña* girl or by noun phrases, e.g. *urqu allqu* male dog, *china
allqu* female dog. Most of the kinship forms distinguish sex of both persons involved in
the relationship:

man's brother	*wawqi*	woman's brother	*turi.*
man's sister	*pani*	woman's sister	*ñaña.*

3.2   Properties of the Noun Dictionary

According to their linguistic attributes words integrate different hierarchical ontology
classes and subclasses. For many entries we have also provided syntactic-semantic
properties. (Remark: most of the examples come from dictionaries QU-FR ou QU-SP
(Itier 2011; Perroud 1970) and, in some cases we give their English translation). For
instance:

1. QU allqu (FR chien) is classified as a common noun (imapa sutin IS, in QU),
   vertebrate animal (wasa tulluyuq quñi yawarniyuq WTQY, in QU) mammal (ñuñuq
   Ñ, in QU).
2. QU pacha (EN cloth) is classified as common noun (IS, in QU), pachakunapaq (PA,
   in QU), clothing, soft thing made of fabric, leather, and so on.
3. QU llaqta (EN city) is classified as a common noun(imapas sutin IS, in QU), proper
   name (runa kaqlla sutiyuq RKS in QU) denoting a geographic (allpaman qatiq, AQ
   in QU) place, geographical entity, and geographical location.

Let us comment the properties accompanying some entries of the sample presented
in Fig. 2.

```
# NooJ V5
# Language is: qu
allqu,N+FR="chien"+FLX=NVOCAL+DRV=YAY+DRV=CHAY+NC+MAMIPH
ayni,N+FR="coopération"+FLX=NVOCAL+DRV=YY+NC
chakra,N+FR="terrain agricole"+FLX=NVOCAL+DRV=YAY+NC+GEO
Inka ñan,N+FR="chemin de l'Inca"+FLX=NCONSO+NPROP+UNAMB
kullu,N+FR="tronc"+FLX=NVOCAL+DRV=YAY+NC
laiqa,N+FR="sorcier"+FLX=NVOCAL+DRV=YY+DRV=YAY+NC
llaqta,N+FR="ville"+FLX=NVOCAL+DRV=YY+DRV=CHAY+DRV=YAY+NC+GEO
mama,N+FR="mère"+FLX=NVOCAL+DRV=YAY+NC
maqta,NM+FR="garçon"+FLX=NVOCAL+DRV=YAY+NC
ñan,N+FR="chemin"+FLX=NCONSO+DRV=YAY+DRV=CHAY+NC
pacha,N+FR="la terre"+FLX=NVOCAL+DRV=YAY+NC+GEO
pacha,N+FR="le temps"+FLX=NVOCAL+NC+TP
pacha,N+FR="vêtement"+FLX=NVOCAL+DRV=YAY+NC+VMT
paukar,N+FR="jardin"+FLX=NCONSO+DRV=YAY+DRV=CHAY+NC
purun,NA+FR="désert"+FLX=NCONSO+DRV=CHAY+DRV=YAY+NC
quillur,N+FR="étoile"+FLX=NCONSO+NC
sipas,N+FR="jeune fille"+FLX=NCONSO+DRV=YAY+NC
waman,N+FR="faucon"+FLX=NCONSO+NC+VTBRSANGCH+AVE
wasi,N+FR="maison"+FLX=NVOCAL+DRV=YAY+DRV=CHAY+NC
yupi,N+FR="trace"+FLX=NVOCAL+DRV=CHAY+NC
yura,N+FR="arbuste"+FLX=NVOCAL+DRV=CHAY+NC
```

Fig. 2. A sample of the dictionary of nouns

1. *allqu* is marked (N) for noun; it inflects according to the morphological paradigm NVOCAL (FLX=NVOCAL) for nouns ending in a vowel, the mark FR for French gives the French translation of the entry: "chien" (dog). We introduce some semantic properties as: common name (NC in FR), mammal animal (MAMIF in FR mammifère), and can derivate into verbs by the suffixes -*yay* and -*chay*;

2. *chakra* a noun (N) which inflects according to the morphological paradigm NVOCAL (FLX=NVOCAL) for nouns ending in a vowel, corresponding to the FR noun "terrain agricole" (cultivated field), with semantic properties as common name (NC in FR), a geographical emplacement (GEO in FR), and can derivate into verb by the suffix -*yay*;

3. *Inca ñan* a noun (N) which inflects according to the morphological paradigm NCONSO (FLX=NCONSO) for nouns ending in a consonant, a proper noun (NPROP) corresponding to the FR noun Chemin de l'Inca (Inca Trail), a multi word unit defined as unambiguous UNAMB;

4. *pacha* a noun (N) which inflects according to the morphological paradigm NVO-CAL (FLX=NVOCAL) for nouns ending in a vowel, corresponding to the FR noun "la Terre" (the earth), with semantic properties as common name (NC), a geo-graphical emplacement (GEO), and can derivate into a verb by the suffix *yay*; which differs from the following homonym;

5. *pacha* a noun (N) which inflects according to the morphological paradigm NVO-CAL (FLX=NVOCAL) for nouns ending in a vowel, corresponding to the FR noun "le temps" (time), with semantic properties as common name (NC), concerning the time (TP), but does not derivate; this form has the following third homonym:

6. *pacha* a noun (N) which inflects according to the morphological paradigm NVO-CAL (FLX=NVOCAL) for nouns ending in a vowel, corresponding to the FR noun "vêtement" (cloth), with semantic properties as common name (NC), relative to clothing (VTM), and can derivate into a verb by the suffix *yay*.

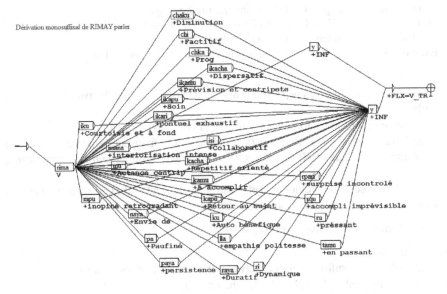

Fig. 3. Graphical grammar for derivation of a transitive verb

Let us note that one QU entry corresponds to one FR transfer, which allows obtaining disambiguated words. For instance, we see that *pacha*=temps (time) and also *pacha*=vêtement (clothing), the first one corresponds to the time (TP) and the second one defines a concrete thing made of fabric or leather or another material which serves to wear. So we try to have the disambiguation at the dictionary level by enriching the lexicon with the addition of syntactic-semantic properties; this is why we try to have as many entries as there are meanings for the same word in the source language. The compiled Nquechua.nod dictionary is able to recognize more than one million inflected forms deriving from 1500 lemmas.

4 The Electronic Dictionary of Quechua Verbs

The electronic dictionary of simple verbs contains **1,472 entries**. It does not include compound verbs, neither phrasal verbs, which are presented in another dictionary. Each verb has an inflectional paradigm assigned to it. For instance the entry unit *rimay* (to talk) inflects according to the paradigm class V_TR, thus the entry becomes: *rimay*,V +FR="parler"+FLX=V_TR whose structure is relatively complex as we will see.

V_TR contains among other things, the paradigms of conjugation in present and future tenses: PR and FUT:

PR=<B>(ni/PR+s+1|nki/PR+s+2|n/PR+s+3|nchik/PR+pin+1|nkichik/PR+p+2
|nku/PR+p+3 |niku/PR+pex+1);
Example: *taki-ni* I sing, *taki-nki* you sing *taki-n* he sings
FUT = <B>(saq/F+s+1 | nki/F+s+2 | nqa/F+s+3 | saqku/F+pex+1 |
sunchik/F+Pin+1 | nkichik/F+p+2 | nqaku/F+p+3);
Example: *taki-saq* I will sing, *taki-nki* you will sing *taki-n* he will sing

It also contains some syntactic and semantic information, like the two main classes of verbs:

Transitive (TR): *rimay* to talk, Intransitive (ITR): *mikuy to eat.*

The intransitive class is quite small. It contains less than one hundred stems. The class of impersonal verbs (INP) includes those relating to weather: *paray* to rain; *lastay* to snow.

As we showed in a previous work (Duran 2013 and 2014), a typical Quechua inflected verbal form has the following structure:

<V><IPS><PR ENDING><PPS>, where:

- V: Verb stem,
- IPS[1]: Interposed suffixes (placed between the lemma and the ending),
- PPS[2]: Post-posed suffix (placed after the ending),
- PR ENDING[3]: is the set of seven present tense personal endings (which will behave as fixed points during the inflections).

Examples:

*rima-**nki***, 'you talk', present tense 1+s
*rima-ri-**nki***, 'you start talking', the IPS suffix *-ri* is interposed
*rima-**nki**_man*, 'you should talk', the PPS suffix *-man* is post posed
rima-ri-***nki***-man, 'you should perhaps start talking', the IPS suffix *-ri* and the PPS suffix *-man* are mixed.

In these examples, each class of suffix intervenes in the inflection only once at both sides of the ending. But, the Quechua grammar allows having several layers of IPS and PPS. (For a more detailed explanation on the construction of valid agglutinations of IPS and PPS suffixes and their mixed patterns see Duran 2014).

---

[1] IPS = {chaku, chi, chka, ykacha, ykachi, ykamu, ykapu, ykari, yku, ysi, kacha, kamu, kapu, ku, lla, mpu, mu, na, naya, pa, paya, pti, pu, ra, raya, ri, rpari, rqa, rqu, ru, spa, sqa, stin, tamu, wa}.

[2] PPS = {ch, chaa, chik, chiki, chu(?), chu, chusina, má, man, m, mi, ña, pas, puni, qa, raq,ri, si, s, taq, yá}.

[3] PR ENDING = {-ni, -nki, -n, -nchik, -niku, -nkichik, -nku}.

In fact, considered separately we will have 231 combinations of two PPS, 132 combinations of three PPS and 92 combinations of four valid combinations of PPS. Here are some samples of the corresponding paradigms:

SVPP2= (CHUSINAVMV MANVCHAAV MANVCHIKV   MANVCHIKIV
ANVQAVMANVCHUIV   MANVRAQV   MANVSIV   MANVTAQV
MANVYAVMANVYAAV   ÑAVCHV ÑAVCHAAV ÑAVCHIKIV ÑAVCHUIV
...)
SVPP3 = (MANVCHUSINAVMV MANVÑAVCHV MANVÑAVCHAAV
MANVÑAVCHIKV MANVÑAVCHIKIV MANVÑAVCHUIV MANVÑAVCHUNV
MANVÑAVCHUSINAV RAQVTAQVYAAV   TAQVCHUSINAVMV...)

On the other hand, a very large number of inflected forms include a mixture of one or more layers of IPS and one or more PPS, we call them mixed inflections MIFLX.

For instance, in the form *rima-ri-lla-**nki**-man-raq* 'I think you should talk before', we have two suffixes IPS (*-ri*, *-lla*) and two PPS (*-man*, *-raq*). Besides, in the form *miku -cha-ku-na-lla-n-paq* «hoping that she will kindly eat it », we have four suffixes IPS (cha, ku, na, lla) and one suffix SPP (*paq*) (4 + 1).

In order to generate automatically all these forms, we have constructed a series of mixed paradigms which we symbolize by V_MIXn. We present the details for V_MIX1 which contains 2*2 = 4 sub-grammars (2 four IPS consisting in one layer, ending in a vowel or a consonant, and 2 for the SPP including one layer placed after the endings, ending in a vowel or a consonant):

    V_MIX1=(:SIP1_PR_V) (:SPP1_V) I (:SIP1_PR_C)(:SPP1_C)
    I (:SIP1_PRM_V) (:SPP1_V) I (:SIP1_PRM_C)(:SPP1_C);

Thus, summarizing, one of the dominant paradigms for the inflection of a verb contains all these grammars:

      V_TR= :V_SPPI:V_TR_SIPI:V_CONJI:IMP; where
          V_PP=:SPP1_C I:SPP1_F I:SPP2;
          V_TR_SIP = :SIP1_GI:SIP2_G;
          SIP1_G = <E>/INFI:SIP1 I:V_SIP1_INFI:SIP1_N; and so on.

4.1   Verb-Verb Derivations: Generating New ULAVs

Among the SIP, we find a sub-set of 27 suffixes which derivate a verb into another verb. This strategy allows Quechua to generate new ULAV[4]. These new forms can be conjugated as if they were simple verbs. We name this set "SIP_DRV".

---

[4] ULAV: 'unités linguistiques atomiques verbales conjugables' or 'Conjugable verbal atomic linguistic units', after Silberztein (2016).

```
# NooJ V5
# Dictionary
rimaruy,rimay,V+FR="parler"+FLX=V_SIP1_PRES_1_réalise_l'action_de_façon_
rimarquy,rimay,V+FR="parler"+FLX=V_SIP1_PAPT_1_accompli_l'action_en_peu_
rimarpariy,rimay,V+FR="parler"+FLX=V_SIP1_ASUR_1_action_surprise_hé:
rimapayay,rimay,V+FR="parler"+FLX=V_SIP1_FREQ_1_répétition_fréquente+INF
rimapay,rimay,V+FR="parler"+FLX=V_SIP1_PEAU_1_peaufine_l'action_PEAU+INF
rimanayay,rimay,V+FR="parler"+FLX=V_SIP1_ENV_1_envié_de,_souhaite_de+INF
rimamuy,rimay,V+FR="parler"+FLX=V_SIP1_ACT_1_actance_centripète_d'approx:
rimakuy,rimay,V+FR="parler"+FLX=V_SIP1_AUBE_1_actance_se_responsabilisan
rimakapuy,rimay,V+FR="parler"+FLX=V_SIP1_RAS_auto_bénéfice_réalise_l'act.
rimakamuy,rimay,V+FR="parler"+FLX=V_SIP1_AAR_aller_àréaliser_l'action+IN:
rimakachay,rimay,V+FR="parler"+FLX=V_SIP1_ARO_action_répété_orienté+INF
rimaisiy,rimay,V+FR="parler"+FLX=V_SIP1_COLL_1_aide_à_réaliser_l'action+:
rimaikuy,rimay,V+FR="parler"+FLX=V_SIP1_COURT_1_courtoisement_soigneusem
rimaikariy,rimay,V+FR="parler"+FLX=V_SIP1_APRP_1_ponctuelle_et_rapidemen·
rimaikapuy,rimay,V+FR="parler"+FLX=V_SIP1_SOIN3_1_avec_attention+INF
rimaikamuy,rimay,V+FR="parler"+FLX=V_SIP1_PREAT_1_vers_le_sujet+INF
rimaikachiy,rimay,V+FR="parler"+FLX=V_SIP1_POLI_1poliment+INF
rimachkay,rimay,V+FR="parler"+FLX=V_SIP1_PROG1_en_train_deréaliser_l'act:
rimachiy,rimay,V+FR="parler"+FLX=V_SIP1_FACT_1_assiste_aide+INF
```

Fig. 4. ULAVs obtained by derivation of the verb *rimay* (to talk)

SIP_DRV= {-*chaku, -chi, -chka, -ykacha, -ykachi, -ykamu, -ykapu, -ykari, -yku,   -
ysi, -kacha, -kamu, -kapu, -ku, -lla, -mpu,-mu, -naya, -pa, -paya, -pu, -raya, -ri, -
rpari, -rqu, -ru, -tamu*}

For example, the NooJ grammar of Fig. 3. generates all the new ULAVs for the
verb *rimay* shown in Fig. 4.

We can obtain combinations of two, three or four IPS suffixes. We may obtain in
this way more than 2,000 derived forms for a single verb. They look like the ones
appearing in the following sample which includes different layers.

*rimarquy,rimay,*V+FR="parler"+FLX=V_SIP_INF+PAPT+INF (talk thoroughly)
*rimarpariy,rimay,*V+FR="parler"+FLX=V_SIP_INF+ASUR+INF (talk to each one with details)
*rimapayaikachaikamuy,rimay,*V+FR="parler"+FLX=V_SIP3_INF+FREQ+ARO+PREAT +
*riariikapuchkay,rimay,*V+FR="parler"+FLX=V_SIP3_INF+DYN+SOIN3+PROG+INF
*rimaisimullay,rimay,*V+FR="parler"+FLX=V_SIP3_INF +COLL+ACENT+POL1+INF
*rimaisimuchkay,rimay,*V+FR="parler"+FLX=V_SIP3_INF+COLL+ACENT+PROG+INF

It is an important question to know the meaning of these generated forms. Are they
only theoretical forms with no tangible sense or are they used currently in the daily
conversations? Many are currently used and are meaningful for the users but their
translation needs long descriptions. For instance, we have the translation of the entry:

rimariikapuchkay,rimay,V+FR="parler"+FLX=V_SIP3_INF+DYN+SOIN3+PROG+INF as follows: 'a kind demand to start talking in behalf of a third person'.

On the other hand, if we attempt to get an automatic translation based in the meaning of each suffix participating in the combination, we may use the 'meaning value' of each of the suffixes noted in the entry, and put them in a grammatical order as follows:

rima	*ri*	*ikapu*	*chka*	*y*
talk		DYN	SOIN3	PROG	INF

The noted symbols carry on the following meanings[5]:

DYN "the action begins"

SOIN3   "a demand to do something in behalf of a third, with care"

PROG   Progressive (the action is taking place at present)

INF   Infinitive

Thus, 'a demand + to do the action of talking + done at present + executed gently in behalf of a third person + to start now' (should be equivalent to) 'a kind demand to start talking in behalf of a third", which seems to be the case.

The complete translation of the new ULAVs is a real challenge for NLP and MT research in Quechua.

4.2 Nominalization

The nominalization suffixes are placed at the end of the verbal lemma. (For details on their construction, see Duran 2013b) They are generated by the following set of suffixes:

S_NV= {*y, na, q, sqa*}.

As we can see in the following examples, certain verbal stems can be nominalized by more than one of them.

> *tiyana*,<$V,N+Instrumental>
> *tiyaq*,<$V,N+ Agentif >
> *rimana*,<$V,N+Instrumental> (an object for talking: microphone)
> *rimai*,<$V,N+Substantif> (the talk)
> *rimaq*,<$V,N+Agentif> (the speaker)
> *rimasqa*,<$V,N+Participe> (the speech)

These forms can be generated by the grammar shown in Fig. 5:

---

[5] A condensate dictionary of 'enunciation moods' of all the IPS suffixes in form of a table can be found in (Duran 2016).

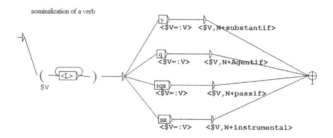

Fig. 5. Nominalization grammar

Gathering all this information concerning the verbs, we have built the dictionary of verbs contained in the Quechua Module for NooJ. We can see in a sample in Fig. 6:

```
# NooJ V5
# Dictionary
# Language is: qu
achay,V+TR+FR="chauffer"+FLX=V_TR+DRV=IS:NCONSO+DRV=QS:NCONSO+DRV=NAS:NVOCAL+DRV=SQAS:NVOCAL
achikyay,V+TR+FR="faire jour"+FLX=V_TR+DRV=IS:NCONSO+DRV=QS:NCONSO+DRV=SQAS:NVOCAL
away,V+TR+FR="tisser"+FLX=IS:NCONSO+DRV=QS:NCONSO+DRV=NAS:NVOCAL+DRV=SQAS:NVOCAL
ayniy,V+TR+FR="aider"+FLX=V_TR+DRV=IS:NCONSO+DRV=QS:NCONSO
aypay,V+TR+FR="atteindre)"+FLX=V_TR+DRV=IS:NCONSO+DRV=QS:NCONSO+DRV=NAS:NVOCAL+DRV=SQAS:NVOCAL
ayuqtay,V+TR+FR="enrôler"+FLX=V_TR+DRV=IS:NCONSO+DRV=QS:NCONSO+DRV=SQAS:NVOCAL
ayuy,V+TR+FR="faire adultère(pour un homme)"+FLX=V_TR+DRV=IS:NCONSO+DRV=QS:NCONSO+DRV=SQAS:NVO
aywiy,V+TR+FR="agiter "+FLX=V_TR+DRV=NAS:NVOCAL+DRV=SQAS:NVOCAL
chakchay,V+TR+FR="mâcher des feuilles de coca"+FLX=V_TR+DRV=IS:NCONSO+DRV=QS:NCONSO
chakiy,V+TR+FR="sécher"+FLX=V_TR+DRV=IS:NCONSO+DRV=QS:NCONSO+DRV=NAS:NVOCAL+DRV=SQAS:NVOCAL
chakmay,V+TR+FR="casser la terre"+FLX=V_TR+DRV=IS:NCONSO+DRV=QS:NCONSO+DRV=SQAS:NVOCAL
chakmay,V+TR+FR="mettre en qachère"+FLX=V_TR+DRV=QS:NCONSO+DRV=NAS:NVOCAL+DRV=SQAS:NVOCAL
chakmay,V+TR+FR="préparer la terre"+FLX=V_TR+DRV=IS:NCONSO+DRV=QS:NCONSO+DRV=NAS:NVOCAL+DRV=SQAS:NVOCAL
challay,V+TR+FR="arroser"+FLX=V_TR+DRV=IS:NCONSO+DRV=QS:NCONSO+DRV=SQAS:NVOCAL
challay,V+TR+FR="asperger "+FLX=V_TR+DRV=IS:NCONSO+DRV=NAS:NVOCAL+DRV=SQAS:NVOCAL
chamqay,V+TR+FR="jeter"+FLX=V_TR+DRV=IS:NCONSO+DRV=QS:NCONSO+DRV=NAS:NVOCAL+DRV=SQAS:NVOCAL
```

Fig. 6. The verb dictionary of the module

5 Dictionaries for Other POSs

We have included, in the module, a dictionary of adjectives and another of other POS containing adverbs, pronouns and numerals. The set of suffixes that allows inflecting each category are given separately. For example the following sets correspond to adjectives and pronouns:

SUF_A_V= {-ch, -chá, -cha,-chik, -chiki, -chu, -chu?, - hina, -kama, -kuna, -lla, -má…, -raq, -ri, -s, -su, -ta, -taq, -wan, -yá, -yupa}, adjectival suffixes.

SUF_PRO_V = {-ch, -chá, ,-chik, -chiki, -chu, -chu?, -hina, -kama, -lla, -má, -man…, -qa, -rayku, -raq, -ri, -s, -ta, -taq, -wan, -yá, -yupa}, pronominal suffixes applied to pronouns ending in vowels.

SUF_PRO_C = (-chá, -chik, -chiki, -chu, -chu?,-hina, -kama, -kuna, -lla, -má, -man, -manta…, -ri, -si, -ta, -taq, -wan, -yupa), pronominal suffixes applied to pronouns ending in consonants.

For instance, here is a sample of paradigm for adjectives ending in a vowel:

A_V_1 = <E>|:CH |:CHAA |:CHIK |:CHIKI |:CHUN |:CHUI |:DCHA|:GEP |:GEPA
|:HINA|:KAMA, ..., |:POSV_v |:POSV_c |:PUNI |:PURA |:QA |:RAYKU |:RAQ |:RI
|:SSIV |:TA |:TAQ |:WAN |:YAA |:YUPA;

This will generate 56 inflected one-dimension forms for the adjective *taksa* (small)
as follows:

> *taksach,taksa*,A+FR=« petit »+ FLX=A_V_1+DPRO
> *taksachá,taksa*,A+FR=« petit »+ FLX=A_V_1+DPRO
> *taksacha,taksa*,A+FR=« petit »+ FLX=A_V_1+DPRO

As for the case of nouns, Quechua allows agglutination of adjectival suffixes up to
seven layers. The following sample shows forms with two suffixes:

> *sumaqkamachá,sumaq*,A+FR="beau"+FLX=A_C_2+LIM+DPRO
> *sumaqkamalla,sumaq*,A+FR="beau"+FLX=A_C_2+LIM+ISO
> *sumaqkunachá,sumaq*,A+FR="beau"+FLX=A_C_2+PLU+DPRO

Applying similar methods to nouns and verbs has driven to hundreds of paradigms,
for the inflection of these categories containing two or more agglutinated suffixes.
Figure 7 shows a sample of the dictionary of pronouns.

```
# NooJ V5
# Dictionary
ñuqa,PRO+FR="je moi"+FLX=FLXPROVOY
qam,PRO+FR="tu toi"+FLX=PROPLU
pai,PRO+FR="il elle lui"+FLX=+FLX=FLXPROCON
pay,PRO+FR="il elle lui"+FLX=FLXPROCON
ñuqanchik,PRO+FR="nous, (tous)"+FLX=PROPLU
ñuqaiku,PRO+FR="nous, (pas vous)")+FLX=PROPLU
kikii,PRO+FR="moi même"+FLX=FLXPROVOY
kikiiki,PRO+FR="toi même"+FLX=FLXPROVOY
kikin,PRO+FR="soi même"+FLX=FLXPROCON
kikinchik,PRO+FR="nous même"+FLX=FLXPROCON
kikiiku,PRO+FR="nous même t'excluant"+FLX=FLXPROVOY
kikiikichik,PRO+FR="vous même pluriel"+FLX=PROPLU
kikinku,PRO+FR="eux même"+FLX=FLXPROVOY
ñoqaiku,PRO+FR="nous autres"+FLX=FLXPROVOY
qamkuna,PRO+FR="vous pluriel"+FLX=FLXPROVOY
juk,PRO+FR="otro"+FLX=FLXPROCON
huk,PRO+FR="otro"+FLX=FLXPROCON
quk,PRO+FR="otro"+FLX=FLXPROCON
chai,PRO+FR="cela"+FLX=FLXPROCON
chay,PRO+FR="otro"+FLX=FLXPROCON
wakin,PRO+FR="quelques uns"+FLX=FLXPROCON
llapan,PRO+FR="tous toutes"+FLX=FLXPROCON
```

Fig. 7. The Dictionary of pronouns

6   Some Additional Grammars to the Library

We present, as an illustration, three graphical grammars contained in the Quechua NooJ Module: Pachayupay.nog (the dates in QU), NumQU.nog (the numbers en QU) and Future.nog (the future in QU).

6.1   The Dates and Numerals in Quechua

Recognition and annotation of the dates is done by the following grammar (Fig. 8):

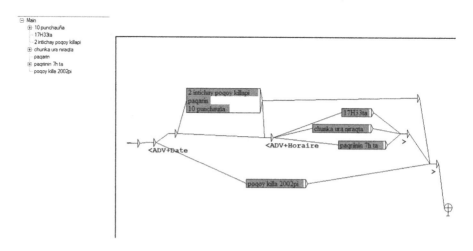

Fig. 8. Pachayupay.nog grammar (the dates in QU)

Figure 9 annotates the numeric determinants in full letters for the numbers going from 1 to 999999: *"huk"* to *"isqun pachak isqun chunka isqunniyuq waranqa isqun pachak isqun chunka isqunniyuq"*.

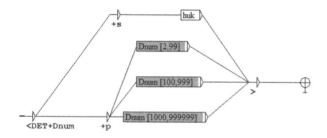

Fig. 9. Annotation of numeric determinants in full letters

6.2   Recognition Grammar for the Future Tense

(See Fig. 10)

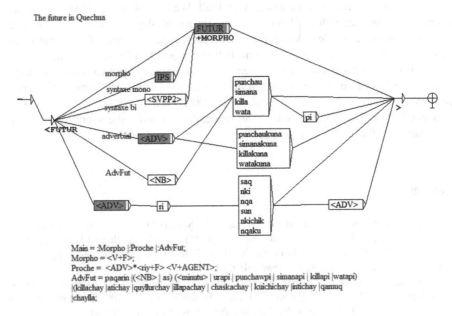

Main = :Morpho |:Proche |:AdvFut;
Morpho = <V+F>;
Proche = <ADV>*<riy+F> <V+AGENT>;
AdvFut = paqarin |(<NB> | as) (<minutu> | urapi| punchawpi | simanapi | killapi |watapi)
|(killachay |atichay |quyllurchay |illapachay | chaskachay | kuichichay |intichay |qamuq
|chaylla;

Fig. 10. Future Quechua.nog grammar (The future in QU)

7   Linguistic Analysis

The linguistic resources that we have presented can be applied to analyze texts. Figure 11 shows an extract of the results of the concordance query <N>+<V+POL1><ADV> applied to a text of Quechua tales. It recognizes and annotates 1,320 forms.

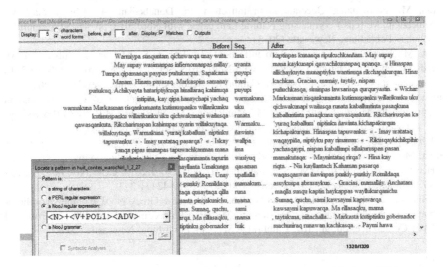

Fig. 11. An extract from the results of the query <N>+<V+POL1><ADV>

8 Conclusion and Perspectives

In this paper we have described the first version of the Quechua Module for NooJ. We have seen how the electronic dictionaries are built. We have shown different grammars and how they interact with the dictionaries and the morphological paradigms. We have also seen a simple example of application of these grammars to a text. The simple verbs dictionary was extracted from M. Duran dictionary Quechua-Français (Duran 2009). Our next step will consist in the enhancement of the French-Quechua verb dictionary, by the inclusion of our current research on the electronic bilingual French - Quechua dictionary containing the translation of the 25,000 French verbal senses of Dubois et Dubois-Charlier 2000. On the other hand, we expect to add some syntactic-semantic grammars and disambiguation grammars.

References

Aoughlis, F., Nait-Serrad, K., Hamid, A., Ferroudja, A., Said, H.M.: New Tamazight module for NooJ. In: Koeva, S., Mesfar, S., Silberztein, M. (eds.) Formalising Natural Languages with NooJ 2013. Selected Papers from the NooJ 2013 International Conference, pp. 13–26. Cambridge Scholars Publishing, Newcastle (2014)

Barreiro, A.: Port4NooJ: Portuguese linguistic module and bilingual resources for machine translation. In: Blanco, X., Silberztein, M. (eds.) Proceedings of the 2007 International NooJ Conference, pp. 19–47. Cambridge Scholars Publishing, Newcastle (2008)

Bogacki, K.: Polish module for NooJ. In: Blanco, X., Silberztein, M. (eds.) Proceedings of the 2007 International NooJ Conference, pp. 48–66. Cambridge Scholars Publishing, Newcastle (2008)

Chadjipapa, E., Papadopoulou, E., Gavrillidou, Z.: New data in the Greek NooJ module: compounds and proper nouns. In: Blanco, X., Silberztein, M. (eds.) Proceedings of the 2007 International NooJ Conference, pp. 96–102. Cambridge Scholars Publishing, Newcastle (2008)

Duran, M.: Dictionnaire Quechua-Français-Quechua, Editions HC, Paris (2009)

Duran, M.: Formalizing Quechua noun inflection. In: Donabédian, A., Khurshudian, V., Silberztein, M. (eds.) Formalising Natural Languages with NooJ2013, pp. 41–50. Cambridge Scholars Publishing, Newcastle (2013)

Duran, M.: Formalizing Quechua verb inflections. In: Koeva, S., Mesfar, S., Silberztein, M. (eds.) Formalising Natural Languages with NooJ 2013. Selected Papers from the NooJ 2013 International Conference, pp. 41–50. Cambridge Scholars Publishing, Newcastle (2014)

Duran, M.: Morphological and syntactic grammars for the recognition of verbal lemmas in Quechua. In: Monti, J., Silberztein, M., Monteleone, M., di Buono, M.P. (eds.) Proceedings of the NooJ 2014 International Conference on Formalising Natural Languages with Noo J2014, pp. 28–36. Cambridge Scholars Publishing, Newcastle (2015)

Duran, M.: The annotation of compound suffixation structure of Quechua verbs. In: Okrut, T., Hetsevich, Y., Silberztein, M., Stanislavenka, H. (eds.) NooJ 2015. CCIS, vol. 607, pp. 29–40. Springer, Heidelberg (2016). doi:10.1007/978-3-319-42471-2_3

Gavrillidou, Z., Chatjipapa, E., Papadopoulou, E.: The new Greek NooJ module: morphosemantic issues. In: Blanco, X., Silberztein, M. (eds.) Proceedings of the 2007 International NooJ Conference, pp. 96–102. Cambridge Scholars Publishing, Newcastle (2008)

Georganta, M., Papadopoulou, E.: Towards an ancient Greek NooJ module. In: Vučković, K., Bekavac, B., Silberztein, M. (eds.) Formalising Natural Languages with NooJ. Selected Papers from the NooJ 2011 International Conference, pp. 41–49. Cambridge Scholars Publishing, Newcastle (2011)

Itier, C.: Dictionnaire Quechua-Français. L'Asiathèque, Paris (2011)

Perroud, P.C.: Diccionario Castellano kechwa, kechwa castellano. Dialecto de Ayacucho. Santa Clara Seminario San Alfonso, Peru (1970)

Silberztein, M.: Formalizing Natural Languages: The NooJ Approach. Wiley, London (2016)

Silberztein, M.: Nooj's dictionaries. In: Vetulani, Z. (ed.) Proceedings of the 2nd Language and Technology Conference. Wydawnictwo Poznańskie Sp. z o.o., Poznan (2005)

NooJ Local Grammars
for Innovative Startup Language

Francesca Esposito and Annibale Elia[(✉)]

Department of Political, Social and Communication Sciences,
University of Salerno, Salerno, Italy
{fraesposito, elia}@unisa.com

Abstract. In this work, we take a linguistic knowledge approach to identify innovative language used to describe new entrepreneurial activity. Starting from the Theory of the Speech Acts, an innovative startup needs to express itself through an innovative language continuously upgraded, to describe processes and products. We use several levels of NooJ to process the type of knowledge presented in business documents to recognize new linguistic resources. This approach, based on Lexicon-Grammar (LG) framework, is extendible to every knowledge domain although we process business documents in Agri-food sector that more than others presents critical issues in knowledge management and representation. Regarding the results of our analysis with NooJ, we note that 10% of tokens retrieved are unknown words.

Keywords: NooJ · Local grammars · NLP applications · Innovative Startup Language · Agri-food startup

1 Introduction

The language used in business is mainly technical and this may turn into a communication gap. Documents are apparently descriptive but not always adequate to circumstances; they have often no truth conditions because they are not properly expressed by who should describe the enterprise. Communication process for innovative entrepreneurial activities is rather complex: they need to describe something which is brand new, that no one has seen or heard before. As Graham [1] explained, a startup is a company designed to scale very quickly. It has its focus on growth unconstrained by geography, which differentiates startups from small businesses. It is necessary for a startup to work on technology, or take venture funding, or have some sort of "exit". Rapid growth is a central element involving business functions, process or products, and the way in which they are described. In recent years, popular lexicon has begun equating startups with tech companies, as though the two are inherently intertwined [2], but the correspondence between the two is not always so strict.

Nevertheless, to be innovative, startups have to commercialize something new that can come out from a technology upgrade, or from the union of many systems or apparently distant disciplines, or from the creation of a new market. This character must determine the expansion of the vocabulary typically used to describe the traditional business functions. That is why the linguistic analysis of innovative startups business

L. Barone et al. (Eds.): NooJ 2016, CCIS 667, pp. 64–73, 2016.
DOI: 10.1007/978-3-319-55002-2_6

documents releases two important questions: in which way the terminology used to describe startups is modified, expanded and updated; and how these updates affect the degree of comprehensibility of the text, impacting on the effectiveness of communication process.

Starting from the Theory of Speech Acts by Austin [3] and successively by Searle [4], we discussed the supply that an NLP environment could give to business communication analysis. Thus, we focused on automatic linguistic analysis of business documents based on Lexicon-Grammar (LG) approach, to detect new terms not yet included in the Italian module of NooJ. Then, we proceeded with the analysis of corpora by using 5 Business Plans of Agri-food startups in digital form. Each corpus presented different words but shared with each other's specific characteristics, particularly in the use of terminology and grammar. We aim to develop a range of rules for the annotation and recognition of entities with specific reference to the terminology used. This approach is extendible to every sector, although, we processed documents in Agri-food sector that, more than others presents critical issues in management and knowledge representation, due to a poor distribution and use of computer applications in the management processes.

2   Improving Business Performances with Automatic Linguistic Analysis

John J. Clancy in 1999 wrote "The invisible power: the language of business" in which he described business as "cultural artifacts", not only as a matter of economics, marketing, and management [5]. Clancy believed in invisible forces (metaphor and other figures of speech) used by leaders to deal more successfully with the economic, cultural, and environmental crises of our times. Communication and more specifically the use of language are critical issues for every aspect of business, particularly when you have to communicate a new process and a new product to market.

Austin [3] affirmed that communication is an act, which can be associated to the identification of the errors that are committed in the use of certain words in everyday reality and an innovative point of access on reality starting from the expressions we use to describe it. Utterances that we consider are not a type of nonsense, but according to the author, they "masquerade" their real meaning.

An utterance has always a meaning, therefore, it is linked to its context of use. By the context in which we use words, we could obtain goals or effects, in addition to the conditions of success. As we show in Table 1, to recognize the performative utterances, from constatives, they are proposed two criteria: one based on grammatical frame and one based on conditions of happiness and unhappiness.

The performative act is a statement that does not describe a certain state of things, does not expose a few facts, but rather allows the speaker to make a real action. Austin would express the sense bound right to an action.

Through a performative act it is fulfilled what you say to do and consequently it produces a real fact. From this discussion, we observed that:

Table 1. Criteria to recognize constative and performative utterances

Grammatical	• Performative verbs (apologize, bet, sort, promise, baptize)	Not all performatives follow this grammatical criterion: *"I will come to your party"*
	• First person present	*"Can you tell me where is Cathedral Square?"*
Conditions of unhappiness/happiness	• Happiness or unhappiness	• Even constatives have conditions of happiness or unhappiness (are also acts)
	• True or false	• If performative do not have truth conditions, may be more or less adequate to the facts

- the clear distinction between constatives and performatives does not exist because every utterance has a performative and constative dimension;
- every utterance has a level of adequacy to reality or circumstances;
- every utterance is an act. As act, it must comply with legitimacy and adequacy rules.

If we consider that every speech act has some effects/goals thus when we formulate a sentence, we could evaluate the effectiveness precisely because of the effects it produces and the goals reached. In particular, this efficacy could be managed considering the hidden meaning contained in every word of statements: they not only describe the action that is taking place, but also achieve it.

In this way, linguistics can help business to improve communication performances. Companies show significant need to communicate, and do it right. Particularly, written business communication, shows a syntactic structure not only more complex than oral, but characterized by a reading "off-line", which is always subject to readers interpretation and therefore probably far from the intention of who wrote it. This often causes misunderstandings: the use of automatic linguistic analysis constitutes a solution that supports even the amount of documents and information that the company produces nowadays.

In the last decades, the interest for the use of language in business has grown: it is recognized that the hidden persuasive linguistic potential improves the positioning of company in the public consciousness. Daniushina observed how the language treatment builds and maintains a good relationship with existing and potential customers, shareholders [6].

Introducing the matter relating to the business language, we must consider two fundamental aspects that make the analysis rather complex. To express the business activities in their complexity as well as in their diversity, we have to consider on the one hand the sublanguages that characterize this world and on the other hand the terminology. For instance, sublanguages are used to describe professional activities belonging to different business sectors: banking, trading, accounting, communication, logistic, administration, and so on. Another issue is referred to terminology: no one could say that business has a specific and limited vocabulary. The study of language in business contexts is highly interdisciplinary [7]. Business activities are so complex that they require the application of several disciplines at the same time and therefore the use of specific languages. Although, it is always necessary that the circumstances, in which terms are uttered, should be in some way, or ways, appropriate. The combination of business functions and processes is impacted by improved communication: from company to company, we have seen language skills consistently deliver tangible business value and virtuous results for organizations that invest in language training.

Ford and Wang [8] observed how the use of language in the field of strategic management has been the subject of many studies [9–12], just because there is no unique classification of words as it exists for other disciplines such as Economy. Every strategic document is a stream of decisions [13] and actions whereby it does not just describe reality but performs it in the same moment in which they are representing it.

3 Methodology

Thus, Natural Language Processing (NLP) used to evaluate business documents of startups, allows to:

1. identify the technological impact in terms of language of a specific innovative business (Innovative Linguistic Resources);
2. recognize the effectiveness of the language used, even verifying its compliance with a real business project.

In our study we analysed the language used and identified new linguistic resources, while we postponed the verify of compliance with real business project to upcoming studies. In order to achieve this project's primary goal with NooJ software [14–19], was analysing business document corpus, it was first necessary to identify the documents. Among business records, Business Plan plays a strategic role in the description of the company: it is subject to evaluation by investors, institutions, providers and offers a complete view of the enterprise. Moreover, it has other important functions: it has to demonstrate clear, concise and precise communication skills; recommend necessary action in future; examine available and possible solutions to a problem, event, situation, or issue; describe business activities; cover the company's situation; analyse business trend and financial activities. Specifically we chose to apply our linguistic analysis to 5Agri-food Business Plan because this sector presents a lack in knowledge management: a fragmentation problem that affects not only businesses, but also the organizations responsible to produce and disseminate knowledge such as universities and research centers. This is also given by a poor distribution and use of computer applications in management [20]. Furthermore, business documents in Agri-food could be rich in technical and terminological words with respective knowledge domains: biological, chemical, medical, gastronomical, economic, and agricultural. Once chosen the type of text to analyse, we proceeded in the following steps:

1. Downloading open business texts
2. Preparing text (text cleaning, normalization)
3. Applying Italian linguistic analysis with NooJ
4. Counting (tokens; Compound words/MWALUs; Unknowns, Domains)
5. Part-Of-Speech Tagging - Innovative Linguistic Resources
6. New Local Grammar

To perform our Linguistic Analysis we have applied Lexicon-Grammar (LG) theoretical and practical framework by Gross [21–25], developed for Italian Module by Annibale Elia, Maurizio Martinelli and Emilio D'Agostino [26]. Applying in further studies [27–32, 33], LG describes the mechanisms of word combinations and provides a complete description of natural language lexical and grammatical structures: lexicon formalization; simple sentences as basic analysis units; election restriction and co-occurrence rules. We proceeded with documents analysis using Italian Module of NooJ. Specifically, we use Linguistic Resources from Italian electronic dictionaries, developed on LG framework, are mainly of two types, separable in the formal and semantic aspect of their content.

- Dictionary of simple words (called DELAS-DELAF) that include all the simple words of Italian, simple word atomic linguistic units (SALUs) and multiword atomic linguistic units (MWALUs).
- Dictionary of compound words (called DELAC-DELACF) defined MWALUs and compound words, which includes the sequences formed of two or more words and which together form each direction unit.

4 Findings

4.1 Linguistic Analysis

In order to process documents we converted every document in MS-Word format and subsequently we unified our 5 Business Plans the in the Text's unit (TU) with a dimension of 71, 4 KB. As predicted, after Linguistic Analysis with NooJ showed in Fig. 1, our corpus presented many new linguistic resources, due to innovative character of business documents analysis. NooJ recognized 5,490 tokens of which 473 unknowns.

Moreover, we observed with NooJ the most recurrent words among all. The first one is offerta (offer); followed by prodotto (product), azienda (company), attività (activities), produzioni (productions), mercato (market), servizi (services), qualità (quality), struttura (building), turismo (tourism). The prevalence of the word "offer" in the text shows us how the Business Plan, as business document, is used not only to describe the company activities but also as a strategic tool of communication and promotion with stakeholders.

4.2 Innovative Linguistic Resources

Among 473 unknown words, about 80 tagged new entries added to our Italian electronic dictionaries: we associated to each entry a morpho-syntactical code, a semantic code, and a inflectional paradigm. Based on their features and properties, we collected Innovative Linguistic Resources dividing them in 4 classes, as follows:

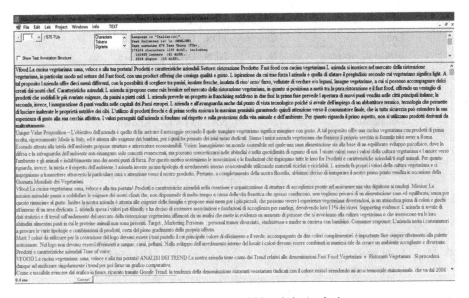

Fig. 1. Text cleaning and Linguistic Analysis

- *Prefixes* (16%). The morphemes that are posted at the beginning of lexemes, are often equipped with an autonomous meaning (agri-campeggio, bio-edilizia, ciclo-turistici).
- *Acronyms* (10%). New forms that express the meaning of MWALUs recently entered the common usage (www, html, ddl).
- *News* (19%). Completely new words, however, are derived from words already present in our linguistic resources (circolareggiante, acrilamide, flavonoidi).
- *Foreign words* (55%). Most words used in business come from the English language, or rather from the American world, and they fit very familiar in our entrepreneurial culture (startup, competitor, asset, follower).

According to the results of our analysis, we built a special dataset with MWALUs and compound words to evaluate our tools. The relation between these linguistic forms and terminology is rather strict. Terminology concerns the unambiguous classification of objects and concepts until to characterize the technical-scientific communication. The terminology cannot be ambiguous and it is therefore in the compounds words and MWALUs that it finds the most appropriate form. In the following Table 2, we collected the most common compound words and MWALUsthat NooJ recognized in the text. For each entry, we indicated: its category; its POS (internal structure); the Domain regarding existent terminological tags in our Italian electronic dictionaries; and the result respect its presence in our Linguistic Resources (known word, unknown word, known but not as compound). Regarding totally unknown words, we noticed the presence of foreign words that mostly described business process.

Table 2. Example of MWALUs and compound words detection

Entry	Category	POS	Domain	Result
Istituti di ricerca	N	NPN	DIGE	Known
Analisidellaconcorrenza	N	NPN	ECON	Known
Costi di gestione	N	NPN	ECON	Known
Capacitàproduttiva	N	NA	ECON	Known
Datistatistici	N	NA	STAT	Known
Economie di scala	N	NPN	ECON	Known
Settoreagricolo	N	NA	ECON	Known
Marketing comparativo	–	–	–	No compounds
Impattoambientale	–	–	–	No compounds
Parco fotovoltaico	–	–	–	No compounds
Business plan	–	–	–	Unknown
Take away	–	–	–	Unknown
Swot analysis	–	–	–	Unknown
Unique value proposition	–	–	–	Unknown

4.3 Domains

As regards the knowledge domains of the entries, we observed that between terminological tags recognized by NooJ there is a prevalence of Economy (63.4%); followed
by Professional (6.8%); Medical (5.9%); Generic Dictionary (5.6%); Engineering
(4.1%); Tourism (3.5%); Law (3.4%); Computer Science (3.1%); Gastronomy (2.5%);
Mathematics (1.3%) (Fig. 2).

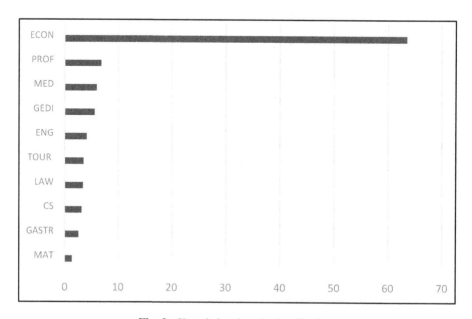

Fig. 2. Knowledge domain classification

4.4   Local Grammars Created for Innovative Startup Language

The Innovative Linguistic Resources have been identified and collected: we added about 80 of these to our Italian electronic dictionaries. In order to build local grammars for our new linguistic resources we generated 6 new inflection codes.

The inflection codes created have matched even new POS structures (AAN) dependent on the fact that more than half of the new linguistic resources are in foreign languages (almost entirely in English). In Fig. 3, we reported an example of compound word local grammar.

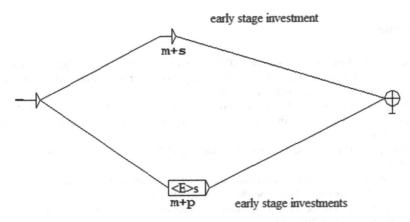

Fig. 3. The graph of inflectional grammar for "early stage investment"

5   Conclusions and Future Works

To be competitive in the market and adapt to the challenges of innovation, startups companies need to acquire specific knowledge, growing and communicating outside their values. Semantic Technologies could help companies to realize these goals: specifically Natural Language Processing applied to business documents could improve the way in which companies communicate. Therefore, the aim of our research was to process the type of knowledge presented in the Business Plans of 5 innovative startups operating in the Agri-food sector. Companies in primary phase of their life express themselves with innovative language not always understandable for stakeholders, and this determinates a communication gap. Then, the Agri-food presents weakness in management and knowledge representation, due to a poor distribution and use of computer applications in management processes.

We used NooJ to create semantic expansion networks, extracting concepts and representing them by means of clustering schemata: about 80 entries have been added to our Italian electronic dictionaries and 6 inflectional codes have been created. We showed how the automatic processing of textual data reducing time spent for measurement and analysis of the project, possibility of a massive control of the documentation, no-dispersion of information related to the company, identifying benchmarks and could monitoring the quality and adequacy of language used. In the future, we propose to identify other Innovative Linguistic Resources: the growth and continue update is the main feature of startups and this determines the expansion of the vocabulary typically used to describe the traditional business functions.

Another issue moves to verify the compliance between the language used in business documents and the real entrepreneurial project and even the effectiveness of language used respect communication goals prefixed. This last point could be realized applying formal description techniques based on conceptualizations and, therefore, on ontologized concepts.

References

1. Graham, P.: Want to start a startup? (2012). http://www.paulgraham.com/growth.html
2. Robehmed, N.: What is a startup? (2013). http://www.forbes.com/sites/natalierobehmed/2013/12/16/what-is-a-startup/#57a38b894c63
3. Austin, J.L.: How to Do Things with Words. Oxford University Press, Oxford (1975)
4. Searle, J.R.: Speech Acts. An Essay in the Philosophy of Language. Cambridge University Press, Cambridge (1969)
5. Clancy, J.J.: The Invisible Powers. The Language of Business. Lexington Books, Lanham (1999)
6. Daniushina, Y.V.: Business linguistics and business discourse. Linguística de negócios e discurso de negócios. Calidoscópio **8**(3), 241–247 (2010)
7. Studer, P.: Linguistics applied to business contexts: an interview with Patrick Studer. ReVEL **11**(21), 187–202 (2013)
8. Ford, E.W., Wang, Z.: Tackling the confusing words of strategy. Effective use of key words for publication impact. Bus. Manag. Strategy **5**(1) (2014)
9. Leontiades, M.: The confusing words of business policy. Acad. Manag. Rev. **7**(1), 45–48 (1982)
10. Hoskisson, R.E., Hitt, M.A., Wan, W.P., Yiu, D.: Theory and research in strategic management. Swings of a pendulum. J. Manag. **25**(3), 417–456 (1999)
11. Nicolai, A.T., Dautwiz, J.M.: Fuzziness in action what consequences has the linguistic ambiguity of the core competence concept for organizational usage? Br. J. Manag. **21**, 874–888 (2010). doi:10.1111/j.1467-8551.2009.00662.x
12. Ronda-Pupo, G.A., Guerras-Martin, L.Á.: Dynamics of the evolution of the strategy concept 1962–2008: a co-word analysis. Strateg. Manag. J. **33**, 162–188 (2012). doi:10.1002/smj.948
13. Mintzberg, H.: Patterns in strategy formation. Manage. Sci. **24**(9), 934–948 (1978)
14. Silberztein, M.: NooJ Manual (2003). www.nooj4nlp.net
15. Silberztein, M.: Corpus linguistics and semantic desambiguation. In: Maiello, G., Pellegrino, R. (eds.) Database, Corpora, Insegnamenti Linguistici, pp. 397–410. Schena Editore/Alain Baudry et Cie, Fasano/Paris (2012)
16. Silberztein, M.: NooJ computational devices. In: Donabédian, A., Khurshudian, V., Silberztein, M. (eds.) Formalising Natural Languages with NooJ, pp. 1–13. Cambridge Scholars Publishing, Newcastle (2013)
17. Silberztein, M.: NooJ V4. In: Koeva, S., Mesfar, S., Silberztein, M. (eds.) Formalising Natural Languages with NooJ 2013, pp. 1–12. Cambridge Scholars Publishing, Newcastle (2013)

18. Silberztein, M.: Analyse et generation transformationnelle avec NooJ. In: Elia, A., Iacobini, C., Voghera, M. (eds.) Livelli di Analisi e Fenomeni di Interfaccia. Bulzoni, Rome (2015)
19. Silberztein, M.: La formalisation des langues: l'approche de NooJ. ISTE Ed., Londres (2015)
20. Ryan, C.D.: Innovation in Agri-food Clusters. Theory and Case Studies. CABI, Wallingford/Boston (2013)
21. Gross, M.: Méthodes en syntaxe. Régime des constructions complétives. Hermann, Paris (1975)
22. Gross, M.: Les bases empiriques de la notion de prédicat sémantique. Langages **15**(63), 7–52 (1981). Larousse, Paris
23. Gross, M.: Lexicon-grammar. The representation of compound words. In: Proceeding of COLING 1986, pp. 1–6. University of Bonn, Bonn (1986)
24. Gross, M.: La construction de dictionnaires électroniques. Ann. Télécommun. **44**(1–2), 4–19 (1989). CNET, Issy-les-Moulineaux/Lannion
25. Elia, A., Martinelli, M., D'Agostino, E.: Lessico e strutture sintattiche. Introduzione alla sintassi del verbo italiano. Liguori Editore, Napoli (1981)
26. Vietri, S.: Dizionari elettronici e grammatiche a stati finiti. Metodi di analisi formale della lingua italiana. Plectica, Salerno (2008)
27. Elia, A., Vietri, S.: Lexis-grammar and semantic web. INFOTEKA **XI**(1), 15–38 (2010)
28. Elia, A., Vietri, S., Postiglione, A., Monteleone, M., Marano, F.: Data mining modular software system. In: Arabnia, H.R., Marsh, A., Solo, A.M.G. (eds.) Proceedings of the 2010 International Conference on Semantic Web & Web Services, SWWS 2010, pp. 127–133. CSREA Press, Las Vegas (2010)
29. Elia, A., Postiglione, A., Monteleone, M., Monti, J.: CATALOGA®: a software for semantic and terminological information retrieval. In: Akerkar, R. (ed.) Proceedings of the International Conference on Web Intelligence, Mining and Semantics, WIMS 2011. ACM, New York (2011). Article 11
30. Elia, A.: On lexical, semantic and syntactic granularity of italian verbs. In: Kakoyianni Doa, F. (ed.) Penser le lexique-grammaire: perspectives actuelles, pp. 277–288. Editions Honoré Champion, Paris (2013)
31. Elia, A.: Operatori, argomenti e il sistema "LEG-Semantic RoleLabelling" dell'italiano. In Mirto, I. (ed.) Relazioni Irresistibili, pp. 105–118. ETS, Pisa (2014)
32. Elia, A., Monteleone, M., Esposito, F.: Dictionnaires électroniques et dictionnaires en ligne. Les Cahiers du Dictionnaire **6**, 43–62 (2014)

Arabic Translation of the French Auxiliary: Using the Platform NooJ

Hajer Cheikhrouhou[1,2(✉)]

[1] University of Sfax, LLTA, Sfax, Tunisia
cheihkrouhou.hager@gmail.com
[2] University of Franche-Comté, ELLIADD, Besançon, France

Abstract. Based on the dictionary of French verbs LVF of Jean Dubois and Françoise Dubois-Charlier, we propose to study the class of auxiliaries X. In this article, we try to translate the verbs of this class into Arabic while comparing the way French and Arabic verbal systems operate. We are going to integrate the verbs of the auxiliary class X (212 entries) into NooJ platform for an automatic translation and we will discuss the similarities as well as the differences between both verbal systems.

Keywords: LVF · Auxiliary class X · NooJ · French and arabic verbal systems · Automatic translation

1 Introduction

Since Aristotle, the verb is associated with tense; it is «Vox significans cun tempore», a phrase which signifies «with tense». The verb is known by its form variations. It is a variable word which is conjugated and thus it signifies different meanings and combines with many markers as number, person, mode, tense and aspect. Mode, tense and aspect constitute three types of classification of verb forms, and which are closely connected and marked by auxiliaries and modal auxiliaries. Indeed, the semantic significance of these grammatical categories is closely related to the enunciative situation i.e. the linguistic utterance which can correspond to three tenses: the past, the present and the future and also to modality which translates the speaker's attitude vis-a-vis the narrated event such as wish, fear, uncertainty or assertion, and so on.

The theoretical framework of this study is based essentially on comparative linguistics which seeks to compare two dissimilar languages, French and Arabic, which have two different linguistic systems and origins.

The present study will be devoted to the linguistic analysis of the grammatical category of French auxiliaries with their Arabic equivalents, within the framework of a comparative contrastive approach, to be integrated in the information system NooJ and applied in automatic translation Silberztein (2015). Our study is based on French verbs (LVF) of Jean Dubois and Françoise Dubois-Charlier (1997) and essentially on the class of auxiliaries X. First, we will analyse the linguistic characteristics of French auxiliaries. Second, we are going to make a comparative analysis between French auxiliaries and

© Springer International Publishing AG 2016
L. Barone et al. (Eds.): NooJ 2016, CCIS 667, pp. 74–86, 2016.
DOI: 10.1007/978-3-319-55002-2_7

their Arabic equivalents to detect similarities and differences between both verbal systems. Finally, we will integrate the auxiliary class X into the NooJ platform for French-Arabic automatic translation.

2   Linguistic Description of the Auxiliaries Class

According to the classification of French verbs made by Jean Dubois and Françoise Dubois-Charlier, the generic class of auxiliaries X contains 212 verbal entries Dubois and Dubois-Charlier (1997). The two authors of LVF established a classification taking into account the adequacy between syntax and semantic interpretation. Then, we are going to describe the verbs in the class X at the semantic and syntactic levels.

2.1   At the Semantic Level

The generic auxiliary class of the LVF contains 212 entries divided into four verbal syntactic-semantic classes.

X1 (114 entries). Contains auxiliary verbs, semi-auxiliary verbs and modal verbs (auxiliaires, semi-auxiliaires, modaux).

> Example: voir(02), être(01), faire(05), devenir, sembler(04), aller(11), mettre 23(s)...
> (to have, to be, to do, to become, to appear, to go, to begin...).
> Cette affaire est épineuse. This affair is thorny.

X2 (43 entries). Contains impersonal verbs (verbes impersonnels).

> Example: il faut, il advient, il se peut... (it is necessary, it happens, it may ...)
> Il faut qu'il termine son travail maintenant. He must finish his work now.

X3 (13 entries). Contains existential verbs (les verbes d'existence)

> Example: exister qp... (to exist)
> Le malheur n'existe pas, toujours le bonheur.
> Misfortune does not exist, always happiness.

X4 (42 entries). Contains inchoative and resultative verbs (inchoatifs et résultatifs.)

> Example: commencer, finir... (to start, to finish...)
> La fête de la musique commence à 5 heures du soir.
> The music festival begins at 5:00 pm (Fig. 1).

G1 CLASSE

B	C	D	G	H	I	K	L	M	X
MOT	DOMAINE	*DOMAINE en clair*	CLAS	OPERATEUR	SENS	CONJUGAISON	CONSTRUCTIO	DERIVATION	LEXIQUE
aboutir 02	TPS	temps	X1a.5	(qc) [aux] résultatif A+infinitif	se conclure à	2aZ	N3a	--1 -- ---- --	5
aboutir 03	TPS	temps	X1a.4	(#) [aux] résultatif A+infinitif	arriver à	2aZ	N1a	--- -- ---- --	5
achever 01	TPS	temps	X1a.4	(#) [aux] résultatif qc/DE+infinitif	finir, parachever	1jZ	T14b0	-2- -3 ---- --	2
achever 02	TPS	temps	X1a.4	(#) [aux] résultatif qc/DE+infinitif	aboutir à, finir par	1jZ	T14b0	--- -- ---- --	5
acquérir 07 (être)	GRA	grammaire	X2a	(QUE+indicatif) [ê] sûr A qn	être sûr	3jZ	N4a - A40	-1- -- ---- --	5
advenir	TPSt	temps	X2a	(QUE+indicatif) [ê] modal A qn	arriver	3gA	N4a - A40	--- -- ---- --	3
agir 06 (s')	GRA	grammaire	X2a	(DE+infinitif/qc) [ê] question	être question de	2aZ	P4000	--- -- ---- --	5
aller 11	TPS	temps	X1a.4	(#) [aux] futur+infinitif	être sur le point de	1zA	T1500	--- -- ---- --	5
apparaître 03	GRA	grammaire	X2a	(QUE+indicatif) [ê] évident A qn	se manifester à	5mB	N4a - A40	--- -- ---- --	5
apparaître 05	GRA	grammaire	X2a	(QUE+indicatif) [ê] évident A qn	il semble	5mB	N4a - A40	--- -- ---- --	5
apparoir	DRO	droit, administration	X2a	(QUE+indicatif) [ê] évident A qn	il apparaît, il semble	00A	N4a - A40	--- -- ---- --	4
appartenir 06	GRA	grammaire	X2a	(DE+infinitif) [ê] A qn	revenir	3gZ	N4a	--- -- ---- --	5
arrêter 05	TPS	temps	X1a.4	(#) [aux] résultatif qc/DE+infinitif	cesser de	1bZ	T14b0	--- -D ---- --	5
arriver 08	TPS	temps	X1a.4	(#) [aux] résultatif A+infinitif	finir	1aZ	N1a	--- -- ---- --	5
arriver 09 (en)	TPS	temps	X1a.4	(#) [aux] résultatif A+infinitif	finir par	1aZ	N1a	--- -- ---- --	5
arriver 11	TPS	temps	X2a	(QUE+indicatif) [ê] modal A qn	se produire	1aZ	N4a - A40	--- -- ---- --	5
attaquer 09	TPS	temps	X4a.2	(#) [init] qc PAR qc	débuter	1aZ	T1306 - A1e	--- -D ---- --	5
avérer 02 (s')	GRA	grammaire	X1a.2	(#) [aux] état+attribut	se montrer, manifester	1gZ	P9006	--- -- ---- --	5
avérer 03 (s')	GRA	grammaire	X2a	(QUE+indicatif) [ê] vrai	il se trouve que	1gZ	P4000	-1- -- ---- --	5
avoir 02	GRA	grammaire	X1a.1	(#) [aux+pp]	accompli	9aZ	A90	--- -- ---- --	5

LVF+

Fig. 1. Extract from the X class

In the dictionary of the French verbs LVF, each verbal entry contains 10 properties: domain, sense, semantic operator, conjugation and derivation…

2.2 At the Syntactic Level

The auxiliaries can have four syntactic constructions: direct transitive (T), an indirect transitive (N), intransitive (A) and pronominal (P).

Direct transitive (T)

Characterize the semi-auxiliaries with a temporal, aspectual or modal value and which the subject may be human or non-human.

The subject is a human (Sujet humain) [T1500].
Example: Pouvoir 01 (can).   Il peut utiliser cette machine.
He can use this machine.
Modal value: have the opportunity to (avoir la possibilité de)

The subject inanimate (Sujet non-animé) [T3500].
Example: devoir 05 (have to).   Cette crise doit être terminée.
This crisis has to be finished.
Modal value: the necessity (la nécessité)

Indirect transitive (N)
The subject is human [N1a].
Example: réussir 05 (to succeed).   Il réussit à terminer ce tableau.
He succeeded to finish this painting.

The subject is human / inanimate [N9a].
Example: commencer 02 (to *start*).   La machine commence à faire du bruit.
The machine starts making noise.
(Subject is inanimate)
Il commence à bavarder.
He starts to talk. (Subject is human)

Intransitive (A)
The subject is human or inanimate [A96].
Example: paraître 10 (to look).   Elle paraît jeune.
She looks young.
Son discours paraît sincère.
His speech seems sincere.

The subject is an inanimate [A30].
Example: débuter 01 (to start).   L'émission débute à midi.
The show starts at noon.

Pronominal (P)
The subject is Completive (que P) and postponed in the verb [P4000].
Example: avérer 03(s) (to turn out).   Il s'avère que ce tableau était une copie.
It turns out that this painting was a copy.

The subject is an inanimate [P30a0].
Example: ramener 15(s) (to bring bak).   Son discours se ramène à nous demander notre aide.
His speech aims at asking for our help.

After presenting the semantic-syntactic properties of the auxiliary class, we will deal with the translation of these verbs into Arabic and we will verify whether French and Arabic verbal systems share the same syntactic and semantic characteristics.

3   The Translation of French Auxiliaries into Arabic

Each translation process engenders similarities and especially differences between two languages from the same origin. These differences become more and more important while dealing with languages of different origins, as in our study, French as an indo-European language and Arabic as a Semitic language. In this context, we will focus on certain cases of translation of French auxiliaries into Arabic. The Arabic equivalents of French auxiliaries can be:

3.1   Verb

Example: commencer 06 (to begin): Meaning = débuter, être au début (to start, be the first). Les négociations commencent entre les deux pays.

بدأت المفاوضات بين البلدين. Negotiations began between the two countries.
Aspectual value: inchoative.

3.2   Tense

Example: avoir 02 (to have): Meaning = accompli (aux + pp) accomplished
J'avais discuté cette idée avec le directeur.

كنتُ قد ناقشتُ هذه الفكرةَ مع المدير . I had discussed this idea with the director.

In Arabic, the plus-que- parfait (past perfect) is translated into
«kāna qad faʻala: كان قد فعل» + temporal value: the anteriority (antériorité) Chairet (1996).

- *Kāna* used to mark the precedence.
- *Qad* expresses modal and aspectual value.
- *Faʻala* is the verbal suffix which marks the anticipation and the accomplished appearance.

3.3   Nominal Sentence

Example: être 01 (to be): Meaning = copule (copula)
Elles sont jolies. هُنّ جميلاتٌ. They are beautiful.
copule Attribut du sujet prédicat: thème
cas: Nominatif

3.4 Multiple Translation

A single predicate can have several verbal entries which generate different translations.
Example: trouver (to find)

1. *Trouver 14 (to find):* Meaning = réussir à (to succeed in), +AR «تَوَفَّقَ».
   Enfin, il *a trouvé* à payer ses dettes.
   أخيرا, تَوَفَّقَ في خلاص ديونه. Finally, he managed to pay his debts.
   Aspectual value: resultative.
2. *Trouver 15(s):* Meaning = exister, y avoir (to exist, exist) +AR= «يوجد».
   Il *se trouve* un magasin tout près. يُوجد مغازةٌ قريبة. There is a shop nearby.
3. *Trouver 17(s):* Meaning = être dans tel état (be in such a state)
   Elle *se trouve* toujours *fatiguée*. هي دائما مُتعبة.
   She is always tired. Adjectif + Cas nominatif

From this comparative study between the two verbal systems, we notice differences at many levels:

Lexical divergence. The auxiliary can be expressed by a morphological or syntactic form different from that of the target language.

Structural divergence. The structure of the French sentence can change in Arabic.

   Example: Il est intelligent. : Sentence (phrase verbale)
   هو ذكي   : Noun phrase (phrase nominale)

Temporal divergence. The French verbal system is richer than the Arabic one which has only three tenses: past, present and future. In fact, it cannot express complex tenses (plus-que-parfait, conditionnel passé). To translate such forms, we add other verbal elements to mark this temporal variable like Abi Aad (2001) kāna:كان, qad:قد.

   Example: plus-que-parfait : kāna qad fa □ ala: كان قد فعل
   Il était arrivé.   Kāna qad waṣ ala.   كان قد وصل.

   Futur antérieur : Sayakūnu qad fa □ ala: سيكون قد فعل
   Il sera arrivé.   Sayakūnu qad waṣ ala.   سيكون قد وصل.

After this level of analysis and after the translation of French auxiliaries into Arabic, we move to the practice of automatic translation using the NooJ platform. In this step, we will integrate the class of auxiliaries into a bilingual dictionary and also we will create formal grammars to remove any ambiguity in syntactic structures Silberztein (2010).

4   The Creation of a Bilingual Dictionary

To formalize the verbs of class X, we start by integrating verbal entries in a bilingual dictionary French-Arabic «**auxiliairedic.dic**» (Fig. 2).

```
#
# Special Features: +NW (non-word) +FXC (frozen expression component) +UNAMB (unambiguous lexical
#                   +FLX= (inflectional paradigm) +DRV= (derivational paradigm)
#
# Special Characters: '\' '"' ' ' ',' '+' '-' '#'
#
#use paraderivational.nof
#use paraflexional.nof
#use paraflexion.nof

aboutir,V+AUXI+Emploi=03+AUX=AVOIR+FLX=FINIR+CONS=N1a+N0VPREPN1+N0Hum+VN+N1Vinf+N1Qph+PREP="à"+DOM
achever,V+AUXI+Emploi=01+AUX=AVOIR+FLX=PESER+CONS=T14b0+N0VN1+N0Hum+V+N1Inanim+N1Vinf+PREP="de"+DR
achever,V+AUXI+Emploi=02+AUX=AVOIR+FLX=PESER+CONS=T14b0+N0VN1+N0Hum+V+N1Inanim+N1Vinf+PREP="de"+DO
acquérir,V+AUXI+Emploi=07+AUX=ETRE+FLX=ACQUERIR+CONS=N4a+N0ETREVPREPN1+N0Imper+ETRE+V+N1Vinf+N1Inan
advenir,V+AUXI+AUX=ETRE+FLX=TENIR+CONS=N4a+N0VPREPN1+N0Imper+V+N1Vinf+N1Inanim+N1Hum+PREP="à"+CONS
agir,V+AUXI+Emploi=06+Emploi=Vpronominal+AUX=ETRE+FLX=FINIR+CONS=P4000+N0seVN1+N0Imper+V+N1Qph+DOM
aller,V+AUXI+Emploi=11+AUX=ETRE+FLX=ALLER+CONS=T1500+N0VN1+N0Hum+V+N1Qph+N1Vinf+DOM=TPS+CLASS=X1a+
apparaître,V+AUXI+Emploi=03+AUX=AVOIR+FLX=CONNAITRE+CONS=N4a+N0VPREPN1+N0Imper+V+N1Vinf+N1Inanim+N
apparaître,V+AUXI+Emploi=05+AUX=AVOIR+FLX=CONNAITRE+CONS=N4a+N0VPREPN1+N0Imper+V+N1Vinf+N1Inanim+N
apparoir,V+AUXI+AUX=ETRE+FLX=APPAROIR+CONS=N4a+N0VPREPN1+N0Imper+V+N1Vinf+N1Inanim+N1Hum+PREP="à"+
appartenir,V+AUXI+Emploi=06+AUX=AVOIR+FLX=TENIR+CONS=N4a+N0VPREPN1+N0Imper+V+N1Vinf+N1Inanim+N1Hum
arrêter,V+AUXI+Emploi=05+AUX=AVOIR+FLX=AIMER+CONS=T14b0+N0VN1+N0Hum+V+N1Inanim+N1Vinf+PREP="de"+DR
arriver,V+AUXI+Emploi=08+AUX=AVOIR+FLX=AIMER+CONS=N1a+N0VPREPN1+N0Hum+VN+N1Vinf+N1Qph+PREP="à"+DOM
arriver,V+AUXI+Emploi=09+AUX=AVOIR+FLX=AIMER+CONS=N1a+N0ENVPREPN1+N0Hum+EN+VN+N1Vinf+N1Qph+PREP="à
```

Fig. 2. Extract of the dictionary of the French auxiliary

Example:

demeurer,V+AUXI+Emploi=03+AUX=AVOIR+FLX=AIMER
+CONS=A96+N0V+N0Hum+V+NMod+DOM=TPS+CLASS=X1a
+OPER="auxétat+attribut"+SENS="rester tel"+LEXI=2+AR= "ظَلَّ "

+Cons=A96 a syntactic construction is intransitive where the subject is human or inanimate followed by a complement to modality. This verb has the sense of «rest-ertel», in Arabic it is translated by "ظَلَّ".

Example: Il **demeure** silencieux. ظَلَّ صَامِتًا He remains silent.

While applying this dictionary to the corpus "Le Monde" using the query <V+AUXI>, we found these results: reste, doit, trouve, pourrait, est, and so on (Fig. 3).

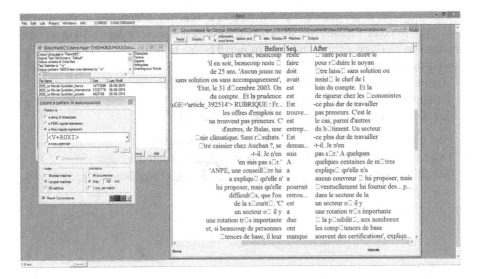

Fig. 3. The results of the query <V+AUXI>

5   NooJ Grammar

For reliable automatic translation of auxiliary, we tried to create grammars to clear up ambiguity of the syntactic construction. In fact, the meaning of the verb is related to the type of subject and the complement (if it is human, concrete…) Silberztein (2014). So, we tried to make grammars that take into consideration the type of the subject and the complement. For disambiguation auxiliary, we created NooJ grammars (Fig. 4):

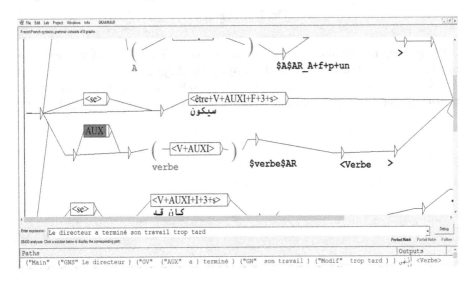

Fig. 4. Grammar 1

By applying the grammar to the sentence «Le directeur a terminé son travail trop tard» (The director finished his work too late), this grammar was able to detect the construction [T1306] of the verb «**terminer 04**» (to finish) whose meaning is «to finish» and translated into «أَنهَى».

Another problem is the translation of the auxiliary in the various tenses.

Example: Avoir(to have), as an auxiliary, is translated into arabic according to its conjugation with the past participle of the verb. Example:

– Plus que parfait (Past perfect) = كان قد فعل (He had done)
– Futur antérieur (Past conditional) = سيكون قد فعل (He will have done)
– Conditionel passé (Present conditional) = يكون قد فعل (Have done) (Fig. 5)

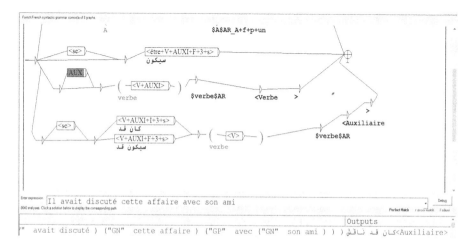

Fig. 5. Grammar 2

By applying this grammar to the sentence «Il avait discuté cette affaire avec son ami» the verb «**avait discuté**» is translated by «كان قد ناقش».

Another issue to resolve is the use of «être» «to be» as a copula followed by an attribute that must be translated by an adjective in the nominative case « ٌ » (Fig. 6).

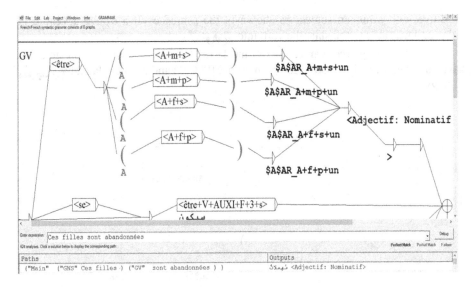

Fig. 6. Grammar 3

The auxiliary «être» followed by the attribute of the subject «abandonnées» is translated by the nominative adjective «مهملاتٌ».

Another problem to be solved is the verbal polysemy. Example: trouver (to find) (Fig. 7).

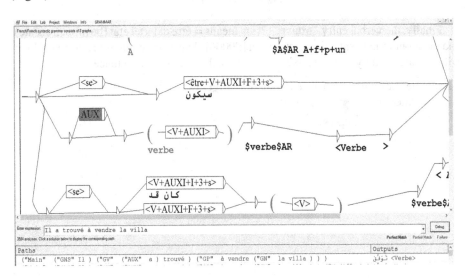

Fig. 7. Grammar 4

The grammar we created was able to detect the meaning of the verb «**trouver 14**» which is «réussir à (succeed in)» and the syntactic construction [N1a]: indirect transitive with a human subject as shown in the sentence «Il a trouvé à vendre la villa» and was able to translate the verb into «توفَّقَ».

The Verbal entry «**trouver 15(s)**» (to find 15 (s)) means = exister, y avoir (to exist, have) and the syntactic construction pronominal [P9001] and it is translated by «يُوجد» (Fig. 8).

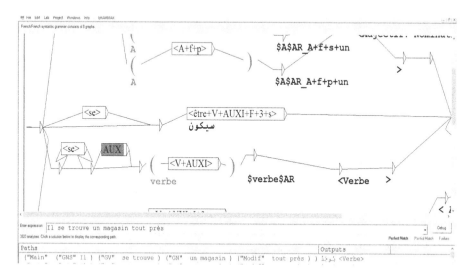

Fig. 8. Grammar 5

Finally, the verbal entry «trouver 17(s)» means = être dans tel état (be in such a state) and the pronominal syntactic construction [P9006]. This verb requires an attribute which will be translated by a nominative adjective used in a nominal sentence.

Example: Les mères se trouvent toujours **fatiguées**.
Is translated by: ↕ (Fig. 9).
<مُتعباتٌ>

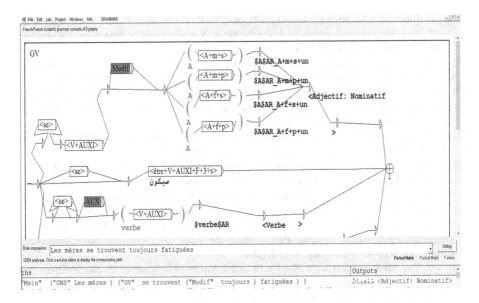

Fig. 9. Grammar 6

6 Conclusion

In this article, we studied the syntactic and semantic characteristics of the auxiliary verbs class X. This analytical study is followed by a comparison between French auxiliaries and their Arabic equivalents showing some syntactic, structural and temporal divergence between both verbal systems. For the application of automatic translation, we integrated French auxiliaries into a bilingual dictionary French- Arabic and we created formal grammars to avoid any ambiguity in syntactic structures and to understand the linguistic differences already found in the analytical and comparative steps. For further research in this fieldwork, we will try to solve the problem of polysemy especially between verbs having the same syntactic construction.

References

Abi Aad, A.: Le système verbal de l'arabe comparé au français: énonciation et pragmatique. Maisonneuve & Larose, Paris (2001)

Chairet, M.: Fonctionnement du système verbal en arabe et en français (Linguistique constractive et traduction). Editions OPHRYS, Paris (1996)

Dubois, J., Dubois-Charlier, F.: Les verbes français. Larousse, Paris (1997)

Le Pesant, D., François, J., Leeman, D.: Présentation de la classification des Verbes Français de Jean Dubois et Françoise Dubois- Charlier. In: Langue Française, 1/2007 (n° 153), pp. 3–19. Larousse, Paris (2007)

Monti, J., Silberztein, M., Monteleone, M., di Buono, M. P.: Formalising Natural Languages with NooJ 2014. In: Selected papers from the NooJ 2014 International Conference. Cambridge Scholar Publishing, Newcastle (2015)

Silberztein, M.: La formalisation du dictionnaire LVF avec NooJ et ses applications pour l'analyse automatique de corpus. In: Langages N°179–180, pp. 221–241. Paris, Larousee (2010)

Silberztein, M.: NooJ v4. In: Koeva, S., Mesfar, S., Silberztein, M. (eds.) Formalising Natural Languages with NooJ 2013. Selected Papers from the NooJ 2013 International Conference, pp. 1–12. Cambridge Scholars Publishing, Newcastle (2014)

Silberztein, M.: Le dictionnaire DEM dans NooJ. In: Bigi, B. (ed.): Actes de TALN 2014, pp. 80–85. Laboratoire Parole et Langage, Aix-en-Provence (2014)

Silberztein, M.: La formalisation des langues: l'approche de NooJ. ISTE Ed., Londres (2015)

Trouilleux, F.: Le DM, a French Dictionary for NooJ. In: Vučković, K., Bekavac, B., Silberztein, M. (eds.) Formalising Natural Languages with NooJ. Selected Papers from the NooJ 2011 International Conference, pp. 16–28. Cambridge Scholars Publishing, Newcastle (2011)

Vučković, K., Bekavac,B., Silberztein, M. (eds.) Automatic Processing of Various Levels of Linguistic Phenomena, Selected Papers from the NooJ 2011 International Conference. Cambridge Scholars Publishing, British (2012)

Wu, M.: The Auxiliary Verbs in NooJ's French-Chinese MT System. In: Donabédian, A., Khurshudian, V., Silberztein, M. (eds.) Formalising Natural Languages with NooJ, pp. 210–222. Cambridge Scholars Publishing, Newcastle (2013)

Corpus Processing and Information Extraction

Integration of a Segmentation Tool for Arabic Corpora in NooJ Platform to Build an Automatic Annotation Tool

Nadia Ghezaiel Hammouda[1(✉)] and Kais Haddar[2]

[1] Miracl Laboratory, Higher Institute of Computer and Communication Technologies of Hammam Sousse, Hammam Sousse, Tunisia
ghezaielnadia.ing@gmail.com
[2] Miracl Laboratory, Faculty of Sciences of Sfax, University of Sfax, Sfax, Tunisia
kais.haddar@yahoo.fr

Abstract. Automatic annotation for Arabic corpora has an important role in many applications of Natural Language Processing (NLP). In this context, we are interested in the automatic annotation of Arabic corpora using transducers set implemented in NooJ platform. And to achieve our aim, we must precede the annotation phase by a segmentation phase. This segmentation phase will, on the one hand, reduce the complexity of the analysis and, on the other hand, improve NooJ platform functionalities. Also, we achieved our annotation phase by identifying different types of lexical ambiguities, and then an appropriate set of rules is proposed. In addition, we experiment our phase on a test corpus with NooJ platform. The obtained results are ambitious and can be improved by adding other rules and heuristics.

1 Introduction

Automatic annotation for an Arabic corpus is currently a challenging research topic in several fields such as translation, syntactic analyses, recognition of named entities and morphological analysis. It is performed on different levels: morphological, lexical and syntactical level. Indeed, this annotation can help in several phases which reduce largely the parsing's execution time. There are several annotated corpora for Arabic language, as the Peen Arabic Treebank (ATB) which is a set of annotated corpora for Modern Standard Arabic (MSA). The ATB annotation encodes the rich morphological structure of words and indicates grammatical relations. Although the corpus annotation stills very expensive in terms of time and money; in fact the ATB corpus is a handcraft work. So any modification or any intervention in its texts requires the repetition of all costly tasks so it will cause a waste of time. As we can see, there is a need for an automatic annotation for corpora, which can be less accurate but less expensive and more time saving. This automatic annotator is based on rich resources, and an effective method of disambiguation. Practically, local grammars can be a useful tool in this disambiguation phase. In fact, finite automata and particularly transducers are increasingly used in NLP. Transduction on text automata is very useful; it can remove paths representing morph-syntactic ambiguities.

© Springer International Publishing AG 2016
L. Barone et al. (Eds.): NooJ 2016, CCIS 667, pp. 89–100, 2016.
DOI: 10.1007/978-3-319-55002-2_8

To release a successful automatic annotation of a corpus, we need to segment it into sentences. In fact, the presence of segmentation leads to reliable results and reduces automatic text processing errors and the execution time. Similarly, having a bad segmentation leads to the accumulation of processing errors. This is why a rigorous study of an Arabic corpus must be done to identify the existing punctuation signs. Also, we need to study the Arabic language to identify lexical rules which can be formalized through different frameworks. To formalize lexical rules, we need to find specific criteria to guarantee a perfect classification. Also, each rule should have a priority allowing an efficient application. The granularity's level of lexical entries can influence the disambiguation phase.

In this context, our objective is to study the Arabic lexical ambiguities and to implement a lexical disambiguation tool for the Arabic language with NooJ platform through the transduction on text automaton. And to facilitate this task, we integrated a segmentation tool for corpora before beginning the disambiguation phase. Also we identified, classified and formalized all Arabic lexical rules (Ghezaiel and Haddar 2015). The obtained NooJ resources will be used to construct an automatic annotation tool.

In this paper, we begin by the state of the art presenting the various previously existing works for the segmentation and the disambiguation phase. Next, we perform our proposed method to resolve lexical ambiguities. Then we implement the segmentation and the disambiguation tools in NooJ platform. These two tools are experimented on a test corpus. Finally, we end the present paper with a conclusion and some perspectives.

2   State of the Art

Many scientific research activities in NLP start their corpus analysis by the choice of a segmentation method. We distinguish two types of methods. The first type is based on punctuation signs and uses symbolic approach (Lungen et al. 2006), (Tofiloski et al. 2009) and (Da Cunha et al. 2010). The second type of segmentation method is based on elementary discourse context using supervised learning approaches (Touir et al. 2008) and (Keskes and Benanamara 2013).

To annotate corpora, some NLP research aim to resolve ambiguities found in different levels: lexical, morphological, syntactical or semantic level. This research use different formalisms. Some works proposed methods for Arabic lexical disambiguation. These methods are based on the cooperation between the morphological analyzer and the syntactic parser like (Shaalan et al. 2004). Also in (Attia 2008), the author developed a parser based on cascade of finite state transducers for sentences pre-processing and sets of syntactic rules for morphological analysis. A third type of disambiguation work dealt with word type classes based on the context linguistic analysis to select the correct sense for a word in a given sentence without resorting to deep morphological syntactic analyses like (Dichy and Alrahabi 2009).

Also, systems exist which deal with Modern Standard Arabic disambiguation. Among these systems we mention MADA and TOKAN's systems (Habash et al. 2009) containing many models. The system AMIRA of (Diab 2009) allows also lexical disambiguation. Besides, we find the Xerox (Beesley 2001) system which is based on finite state technology tools (e.g. xfst, twolc, lexc) allowing the disambiguation's phase.

As well, there are several Arabic works using NooJ platform that address to the disambiguation phase. We cite the work presented in (Fehri et al. 2011) for Arabic named entities translation. Besides, there are many works specialized in one particular phenomenon. Also, we can cite the work described in (Ellouze et al. 2009) to analyze the Arabic broken plural and the work of (Hasni et al. 2009) for elliptical sentence resolution.

As we can see, the Arabic disambiguation phase is not yet well performed because of some difficulties linked to the Arabic's specifies of structure. Lexical rules are not well formalized. Consequently their implementation is not easy. Also, many works in NooJ platform are limited in just one level or one phenomenon. Their reuse is difficult on account of their incompatibility and the absence of consistency.

3   Identification of Lexical Rules Resolving Arabic Ambiguities

In order to identify lexical rules resolving several forms of ambiguities, we carried out a linguistic study. According to this study, Arabic lexical ambiguity can be caused by several aspects. We distinguish four Arabic ambiguity's causes:

Unvocalization. It can cause lexical ambiguities. For example, the word *dhahaba[1]* can refer to noun *the gold*, or the verb *to go* in English.

The emphasis sign (Shadda $\ddot{\circ}$). In Arabic, emphasis sign *Shadda* is equivalent to write the same letter twice. The insertion of *Shadda* can change the meaning of the word. For example, the word *darasun_* means *cours* (noun) while *darrasa* means *he taught* (verb).

Hamza sign. The presence of *Hamza sign* reduces ambiguities. The word *'idh-inun* is a noun that means *permission*. If we delete the *Hamza sign* the obtained word can be the verb *'adhina* that means *authorize*.

Agglutination. In Arabic, particles, prepositions and pronouns can be attached to adjectives, nouns, verbs and particles. This characteristic can generate many types of lexical ambiguity. For example, the character *faa'* in the word *faS-lun (season)* is an original character while in the word *faSa-laa* (then he prayed) is a prefix.

The mechanism of sub-categorization for verbs, nouns and particles help us to construct a set of lexical rules. So, particles can be subdivided into three categories: particle acting on nouns, particles acting on verbs and particles acting on both nouns and verbs. For verbs, transitivity can play a role in the rule's identification. In Arabic, a verb can be intransitive, transitive, double transitive and triple transitive. Either transitive or intransitive verbs can be transformed into transitive type with prepositions. Concerning nouns, the contextual and lexical indicators can be used. In fact, the most reliable indicators for the detection and categorization are right and left contexts of a word. These indicators are either internal or external. Internal indicators are located inside the named entity while external indicators refer to the existing of entity's occurrence.

4   Proposed Method

Our proposed approach consists of three main phases: the segmentation, the preprocessing and the disambiguation phases. The segmentation phase consists of the identification of sentences based on punctuation signs illustrated in Table 1.

Table 1.   Classification of punctuation's markers

Type	Symbols	Position
Point		At the end of the sentence with complete sense
Semicolon	؛	Between two sentences whose the second will cause the first
Two points	:	Explanation: between words Citation/Dialog
Exclamation mark	! !! !؟	End of exclamatory sentence
Interrogation mark	؟ ؟؟ ؟!	End of the interrogative sentence

Table 1 describes the proposed classification of the punctuation's markers. In fact, each marker has a special position in Arabic sentence. The point sign can be at the end of the well formed sentence. For the semicolon, it can be placed between two sentences whose the second will cause the first. But for the two points, they can be placed in the middle in case of explanation or dialog. And for the exclamation mark or the interrogation mark, it can be placed at the end of an exclamatory sentence or an interrogative sentence. We have adopted this classification after a rigorous study of our corpus of study. Then we established transducers in NooJ platform representing all extracted rules that represent our classification of punctuation's markers.

In our segmentation phase, each identified sentence is delimited by an XML tag representing the output of this phase. And as a final result, we obtain an XML document containing a well segmented corpus which will be the input for the preprocessing phase. The second phase consists of the Arabic agglutination's resolution using morphological grammars. As an output of this phase, we obtain a Text Annotation Structure (TAS) containing all possible annotations for corpus's sentences. The obtained TAS is the input of the third phase. The last phase aims to identify the adequate lexical category of each word in given sentence. This identification is based on syntactic grammars specified with NooJ transducers. Transducer's applications respect a certain priority (Fig. 1). So the high priority is given to the most evident transducer. The output of the disambiguation phase will be a disambiguated TAS containing right paths and right annotations. In our method, we used a high granularity's level for lexical categories. Indeed, we have distinguished between nominative, accusative and genitive modes for nouns.

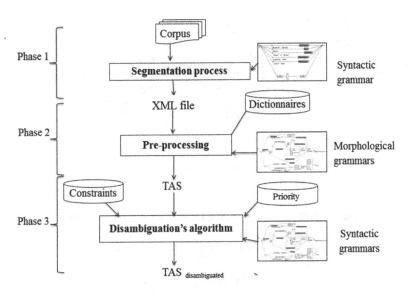

Fig. 1. Proposed method

5   Implementation Phase

In the following, we present the implementation of the already presented phases.

5.1   Segmentation Phase

The implementation of the segmentation phase is based on a set of the developed transducers in NooJ linguistic platform. This set contains 9 graphs representing contextual rules. The main transducer presented in Fig. 2 adds an xml tags <S> to delimit the frontiers of a sentence.

Fig. 2. The main transducer for XML markup

This transducer contains two subgraphs. The first called "Sentence" recognizes a sentence composed only from a sequence of words, numbers and tabulations. Figure 3 illustrates this subgraph.

The second subgraph called "SPunctuation" recognizes the punctuation signs of Table 1. This subgraph is presented in Fig. 4.

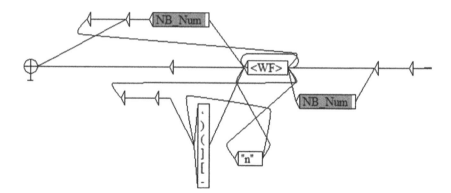

Fig. 3. Transducer recognizing a sentence without punctuation

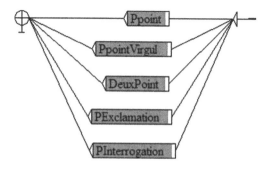

Fig. 4. Transducer recognizing punctuation signs

Figure 4 shows that five states exist according to identified punctuation signs.

5.2 Preprocessing Phase

The implementation of the preprocessing phase is based on a set of morphological grammars and dictionaries (Mesfar, 2008) existing in NooJ linguistic platform. This implementation resolves all forms of agglutination. The outputs contain all possible lexical categories of each word in sentences.

5.3 Disambiguation Phase

The identified lexical rules have been specified with NooJ transducers. The implementation of this phase is based on 17 local grammars representing lexical and contextual rules.

Note that, the Arabic sentence can be either verbal or nominal. So, we construct transducers to recognize these two specific forms. A nominal sentence is generally formed by a topic and an attribute. Figure 5 illustrates a transducer recognizing a nominal sentence.

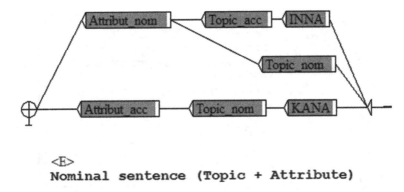

<E>
Nominal sentence (Topic + Attribute)

Fig. 5. Transducer representing a lexical rule for a nominal sentence

Transducer of Fig. 5 takes into account the granularity. Indeed, a nominal sentence contains a nominative topic followed by a nominative attribute. However, after a modal verb "KANA" we find a nominative topic followed by an accusative attribute and the modal verb "INNA" is followed by an accusative topic and a nominative attribute.

6   Experimentation and Evaluation

To experiment our proposed method, we used a test corpus that contains 20000 meaningful sentences mainly from Tunisian newspapers and children's stories. Also, we used dictionaries containing 24732 nouns, 10375 verbs and 1234 particles.

After applying the segmentation tool to the test corpus, a modified corpus containing new XML tags is obtained. And to generate the final output in this phase, the TAS is exported to NooJ option: *Text→ export annotated text as XML document.*

Figure 6 represents an extract from the generated XML document. To include this document into the preprocessing file, we import it as a corpus with XML annotations using the NooJ option: Text→ Open→ doc.xml (Fig. 7).

<S>حينما ذبلت الوردة الطبيعية، شمتّ بها الوردة الصناعية، وقالت ساخرة:<S/>
-<S>أرأيت قد ذبلت سريعا!<S/>
-<S>إذا فبارقة أردني، لا أعيش إلا قليلا.<S/>
-<S>أنت تأعيش عمرا طويلا.<S/>
-<S>طول العمر، لا يدعو إلى الفخر.<S/>
-<S>وبأي شيء تفخرين؟<S/>
-<S>بما تجعلني للآخرين.<S/>
-<S>وماذا أعطيت في عمرك القصير؟<S/>
-<S>أعطيت البهجة والعطر، فهل أعطيت أنت شيئا في عمرك الطويل؟<S/>

Fig. 6. Extract from XML document

Figure 7 shows the selection of the option "XML Text Nodes" and the insertion of the new text unit delimiter in the XML document. In our case, the delimiter is <S>. An XML text will be opened in NooJ platform as the input for the preprocessing phase. All treatments will be done only on units delimited by the opened tag <S> and closed tag </S>.

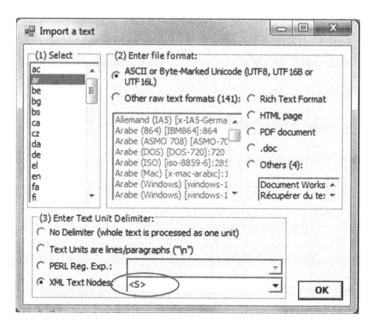

Fig. 7. Import the corpus with XML annotations

Besides, we used in our experimentation a list of morphological grammars containing 113 inflected verb form patterns, 10 broken plural patterns and 3 agglutination grammars. So, we used 17 graphs representing lexical rules, and a set of 10 constraints describing the execution of rules application. The obtained result is illustrated in Table 2.

Table 2. The obtained result of the disambiguation phase

Corpus	Number	Percentage
Sentences	20000	100%
Totally disambiguated	12000	60%
Partial disambiguated	6000	30%
Failed disambiguation	2000	10%

Table 2 shows that 12000 sentences from the 20000 sentences existing in the corpus were totally disambiguated, which represents 60%. Also, there are 6000 sentences partially disambiguated, which represents 30%, and only 2000 sentences erroneously disambiguated, which represents 10%. The partial disambiguation is due to the lack of semantic rules. Also, sometimes, some rules were not correctly recognized. The erroneous disambiguation is linked to the lack of some information in our dictionaries which led to the wrong detection of left or right contexts.

There is another reason for this partial disambiguation which is granularity effect. In fact, the POS tagging can be perfect, if the granularity is characterized by a high level: containing different morphological, lexical and syntactic information as declination, mode and gender. Certain features characterizing lexical categories are illustrated in Table 3. Also, the annotation can be improved, if we resolve the inconsistencies related to certain characteristics of the Arabic language.

Table 3. Features characterizing lexical categories

	Noun	Verb	Pronoun
Type	Noun : NUM, PROP, Adjective : COMP, NUM,	I, C, P	POSS, REL, DEM
Function	–	SUBJ	–
Mode	NOM, ACC, GEN	–	–
Definition	DEF, INDEF	–	–
Agreement	MASC, FEM SG DU PL	M, F S, D, P 1, 2, 3	M, F S, D, P 1, 2, 3

Now if we decrease the granularity, several subsets of categories will be factored into a single category. As an example, the subset {V:I, V:I+SUBJ, V:I+SUBJ+M, V:I +SUBJ+M+S, V:I+SUBJ+M+S+2} which indicates the verb with three labels in the case of high granularity. If we have a low granularity, labels are generalized and factored into a single label under the Verb V. This is explained by the fact that the lack of precision in the grammatical categories (due to the decrease in the level of granularity) allows factoring. This factorization influences the number of rules in the disambiguation phase, and therefore its size.

To illustrate the granularity effect, we will apply our set of syntactic grammars with different levels of granularity on the following sentence:

Saafaha al-walada al-ladhi najaha
He greetsthe boy who succeeded

We have defined four granularity levels concerning the POS tagging. Level 0 contains the basic feature (POS), level 1 is level 0 enriched by the agreement information features, and level 2 added the valence feature to level 1. Besides, level 3 is the level 2 enriched by the mode features (nominative, accusative and genitive).

The obtained results are illustrated in the following Table 4.

Table 4. The granularity's results

Granularity's level	Number of paths	Execution's time (s)
Level 0	1275	9.8
Level 1	956	7.6
Level 2	652	5.6
Level 3	480	4.9

Table 4 shows that if the level of granularity increases, the number of paths decreases and the execution's time too.

Thus the change of granularity level can improve our disambiguation's result.

Figure 8 illustrates that the high annotation level is linked to the degradation in the number of paths so the reduction in execution's time.

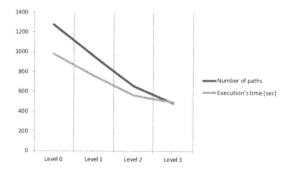

Fig. 8. The variation of granularity over the time

Our evaluation is performed by the precision, recall and the F-measures (Table 5).

Table 5. Values of used measures

Corpus	Precision	Recall	F-measure
20000	0,6	0,9	0,72

Table 5 shows that the obtained values of measures are ambitious. We must improve the precision by adding other rules and heuristics.

7   Conclusion and Future Works

In this paper, we established a segmentation tool for Arabic corpora in the NooJ platform based on a set of punctuation signs extracted from a study corpus. This tool is integrated in a disambiguation one to annotate Arabic corpora automatically. So, we conducted a study on the different types of Arabic lexical ambiguities and syntactical mechanisms to identify the adequate lexical rules. In fact, this study allowed us to establish a set of lexical rules and constraints for Arabic lexical disambiguation. Established rules are specified with NooJ transducers. This disambiguation phase will help us to reduce later the parsing complexity. Thus, an experiment is performed on an Arabic corpus mainly from children tales. The obtained results are satisfactory. This is proven by the calculated measures. Also, we have shown that the use of transducer cascades on the text automata is interesting and effective, can simplify the ambiguity resolution process and make it more effective.

As perspectives and to perfectly annotate corpora, we will enrich our resources by creating new dictionaries and new local grammars representing the maximum of lexical rules. We need also to enrich our set of rules by adding new morphological, syntactic and semantic levels and extend this methodology to other phenomena (i.e., coordination). Also, we want to write new local grammars allowing the improvement of the lexical disambiguation. After, we want to implement a module allowing the information extraction from the annotated corpora. This module can be integrated later in the NooJ linguistic platform.

References

Beesley, K.: Finite-State Morphological Analysis and Generation of Arabic at Xerox Research. Status and Plans (2001). http://www.xrce.xerox.com/content/download/23387/170683/version/1/file/finite-state.pdf

Da Cunha, I., SanJuan, E., Torres-Moreno, J.-M., Lloberes, M., Castellón, I.: Discourse segmentation for Spanish based on shallow parsing. In: Sidorov, G., Hernández Aguirre, A., Reyes García, C.A. (eds.) MICAI 2010. LNCS (LNAI), vol. 6437, pp. 13–23. Springer, Heidelberg (2010). doi:10.1007/978-3-642-16761-4_2

Dichy, J., Alrahabi, M.: Levée d'ambiguité par la methode d'exploration contextuelle: la sequence 'alif-nûn' en arabic. In: Sidhom, S., Ghenima, M., Ouksel, A. (eds.) Information Systems and Economic Intelligence: Proceedings of the 2nd International Conference SIIE 2009, pp. 573–585. IHE Editions, Tunis (2009)

Diab, M.T.: Second generation tools (AMIRA 2.0): fast and robust tokenization, POS tagging, and base phrase chunking. In: Choukri, K., Maegaard, B. (eds.) Proceedings of the Second International Conference on Arabic Language Resources and Tools, pp. 285–288. The MEDAR Consortium, Cairo (2009)

Ellouze, S., Haddar, K., Abdelhamid, A.: Study and analysis of Arabic broken plural with NooJ. In: Ben Hamadou, A., Mesfar, S., Silberztein, M. (eds.) Proceedings of the NooJ 2009 International Conference and Workshop, pp. 31–50. Sfax University Publication Center, Sfax (2010)

Elyes H., Haddar, K., Abdelhamid, A.: Reconnaissance des expressions elliptiques arabes avec NOOJ. In: Proceedings of the 3rd International Conference on Arabic Language Processing, Rabat, Marocco (2009). http://www.citala.org/citala2009/docs/citala%20papers/(N%25B012-Paper%2042).pdf

Fehri, H., Haddar, K., Abdelhamid, A.: Recognition and translation of arabic named entities with NooJ using a new representation model. In: Constant, M., Maletti, A., Savary, A. (eds.) Proceedings of the 9th International Workshop on Finite State Methods and Natural Language Processing, pp. 134–142. ACM, New York (2011)

Ghezaiel, N., Haddar, K.: Study and resolution of Arabic lexical ambiguity through transduction on text automaton. In: Okrut, T., Hetsevich, Y., Silberztein, M., Stanislavenka, H. (eds.) NooJ 2015. CCIS, vol. 607, pp. 123–133. Springer, Heidelberg (2016). doi:10.1007/978-3-319-42471-2_11

Habash, N., Rambow, O., Roth, R.: MADA+TOKAN: a toolkit for Arabic tokenization, diacritization, morphological disambiguation, POS tagging, stemming and lemmatization. In: Choukri, K., Maegaard, B. (eds.) Proceedings of the Second International Conference on Arabic Language Resources and Tools, pp. 102–109. The MEDAR Consortium, Cairo (2009)

Keskes, I., Benanamara, F., Belguith, L.: Segmentation de textes arabes en unites discursives minimales. In: TALN (eds.) Actes de la 20e conférence sur le Traitement Automatique des Langues Naturelles (TALN'2013), pp. 435–449. TALN, Les Sables d'Olonne (2013)

Lungen, H., Lobin, H., Barenfanger, M., Hilbert, M., Puskás, C.: Text parsing of a complex genre. In: Martens, B. (ed.) Proceedings of the Conference on Electronic Publishing, pp. 247–256. ELPUB, Bansko (2006)

Mesfar S.: Analyse morpho-syntaxique automatique et reconnaissance des entités nommées en arabe strandard. Ph.D. thesis, University of Franche-Comté, France (2008)

Shaalan, K., Othman, E., Rafea, A.: Towards resolving ambiguity in understanding Arabic sentence. In: Choukri, K. (ed.) Proceedings of the International Conference on Arabic Language Resources and Tools, pp. 118–122. The MEDAR Consortium, Cairo (2004)

Silberztein, M.: Disambiguation tools for NooJ. In: Varadi, T., Kuti, J., Silberztein, M. (eds.) Applications of Finite-State Language Processing. Selected Papers of the 2008 International NooJ Conference, pp. 1–14. Cambridge Scholar Publishing, Newcastle (2010)

Touir, A., Mathkour, H., Al-Sanea, W.: Semantic-based segmentation of Arabic texts. Inf. Technol. J. **7**(7), 1009–1015 (2008)

Tofiloski, M., Brooke, J., Taboada, M.: A syntactic and lexical-based discourse segmenter. In: Lee, G.G., Shulte, S. (eds.) Proceedings of the ACL-IJCNLP 2009 Conference, pp. 77–80. ACL, Stroudsburg (2009)

Semi-automatic Part-of-Speech Annotating for Belarusian Dictionaries Enrichment in NooJ

Yury Hetsevich[1(✉)], Valery Varanovich[2], Evgenia Kachan[1],
Ivan Reentovich[1], and Stanislau Lysy[1]

[1] United Institute of Informatics Problems, Minsk, Belarus
yury.hetsevich@gmail.com, evgeniakacan@gmail.com,
mwshrewd@gmail.com, stanislau.lysy@gmail.com
[2] Belarusian State University, Minsk, Belarus
gamrat.vvv@gmail.com

Abstract. This paper describes the algorithm for the Belarusian main dictionaries enrichment in NooJ, on the basis of the first one-million corpus for the Belarusian NooJ module. From the broad list of possible subject categories, the corpus focuses on literature of fiction, historical literature, medical literature, scientific literature, sociological literature, and so on. The corpus is considered the finest source for searching unknown words of different domains. So, for this purpose a specific algorithm for automatic word paradigms generation have been agreed to develop. The authors have worked out a mechanism for further processing of all unknown (unique) words extracted from the corpus and adding them to the present dictionary on the basis of the Belarusian NooJ module. The algorithm is based on the required grammatical information of an entire word.

Keywords: Corpora · Belarusian Nooj-Module · Part-of-Speech Tagging · Counter-Check · Lexicology · Dictionary · Algorithm · Online-Service · Paradigm

1 Introduction

This research is a continuation of the overall work on the creation of The First One-million Corpus [1] for the Belarusian NooJ Module [2], which is applicable in a variety of thematic spheres and can be used in any linguistic research.

Today, the first one-million Belarusian corpus for the Belarusian NooJ module is used for solving the following fundamental tasks: optimizing and expanding the development of high-quality linguistic algorithms for the electronic text pre-processing in the TTS (Text-to-Speech) system [6].

The main task of the research is to work out a mechanism for further annotation of different categories and paradigms according to inflection classes of all unknown words extracted from Belarusian corpus and then to compose processed words to main Belarusian NooJ dictionary [3].

© Springer International Publishing AG 2016
L. Barone et al. (Eds.): NooJ 2016, CCIS 667, pp. 101–111, 2016.
DOI: 10.1007/978-3-319-55002-2_9

2 The Part-of-Speech Tagging Countercheck of Unknown Words

The corpus (see Fig. 1) was developed in an appropriate format for Nooj program last year. It composes 338 text files, where the total number of all word forms in the texts is more than 1 million, 197712 of which are unique well-known word forms (received by the <DIC> query, 1 occurrence per match) and 50186 – the unique unknown word forms (received by the <UNK> query, 1 occurrence per match) [1].

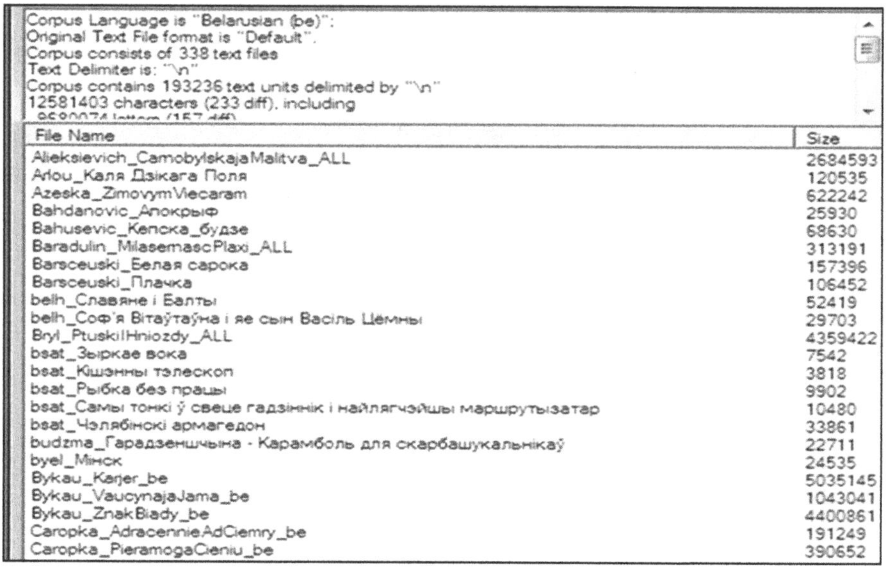

```
Corpus Language is "Belarusian (be)":
Original Text File format is "Default".
Corpus consists of 338 text files
Text Delimiter is: "\n"
Corpus contains 193236 text units delimited by "\n"
12581403 characters (233 diff), including
9500074 letters (157 diff)
```

File Name	Size
Alieksievich_CamobylskajaMalitva_ALL	2684593
Arlou_Каля Дзікага Поля	120535
Azeska_ZimovymViecaram	622242
Bahdanovic_Апокрыф	25930
Bahusevic_Kepcкa_будзе	68630
Baradulin_MilasemascPlaxi_ALL	313191
Barsceuski_Белая сарока	157396
Barsceuski_Плачка	106452
belh_Славяне і Балты	52419
belh_Софʼя Вітаўтаўна і яе сын Васіль Цёмны	29703
Bryl_PtuskilHniozdy_ALL	4359422
bsat_Зыркае вока	7542
bsat_Кішэнны тэлескоп	3818
bsat_Рыбка без працы	9902
bsat_Самы тонкі ў свеце гадзіннік і найлягчэйшы маршрутызатар	10480
bsat_Чэлябінскі армагедон	33861
budzma_Гарадзеншчына - Карамболь для скарбашукальнікаў	22711
byel_Мінск	24535
Bykau_Karjer_be	5035145
Bykau_VaucynajaJama_be	1043041
Bykau_ZnakBiady_be	4400861
Caropka_AdracennieAdCiemry_be	191249
Caropka_PieramogaCieniu_be	390652

Fig. 1. The fragment of the first one-million corpus for the Belarusian NooJ module

Then a specialized dictionary of unknown words was composed for easy determination of categories for these words, firstly automatically, and then it was checked by linguists/experts (see Fig. 2).

During the research of this year, new results to determine the unique categories of unknown words from the first million corpus for the Belarusian NooJ module were received. Statistical data are presented in Table 1 under date of 25.05.2016.

- The total quantity of all unique unknown words after their lowercase conversion and spellcheck is 47206.
- The total quantity of annotated unique unknown words is 26983.
- The total quantity of unique unknown words annotated by the categories NOUN, ADJECTIVE and VERB is 21845.

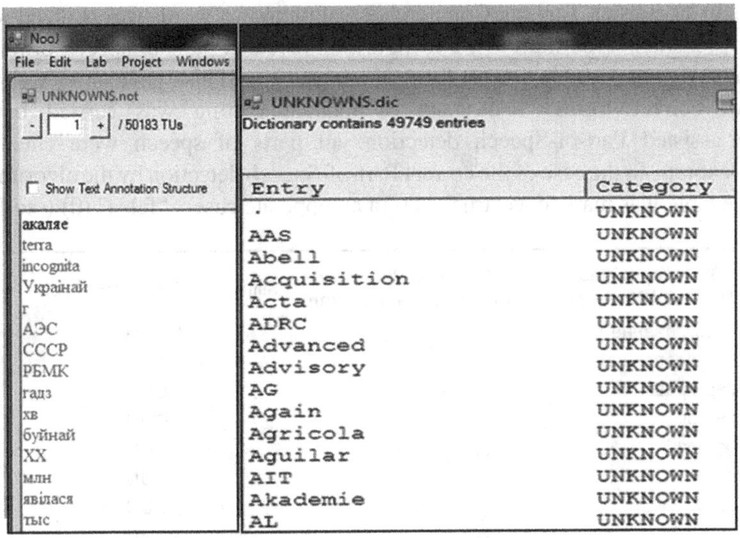

Fig. 2. The fragment of the specialized dictionary of unknown words

- The total quantity of unique unknown words annotated by the remaining categories is 5138.
- The total quantity of unannotated unique unknown words is 18836.

Table 1. The statistics of **POS-annotated** and **POS-unannotated** unique unknown words in the first one-million corpus for the Belarusian NooJ module

Main information about the unknown words	Quantity	Quantity (%)
All unique unknown words (according to NooJ results)	50 183	100,00
All unique unknown words (after their lowercase conversion and spellcheck)	47 206	94,07
The processed part	2 977	5,93
Words annotated by categories: general and additional	Quantity	Quantity (%)
NOUN	12 303	45,60
VERB	4 843	17,95
ADJECTIVE	4 699	17,41
PARTICIPLE	1 495	5,54
FOREIGN	1387	5,14
ADVERB	981	3,64
PRONOUN, NUMERAL, PREPOSITION, CONJUNCTION, PARTICLE, PARENTHESIS, INTERJECTION, PREDICATIVE	553	2,05
ABBREVIATION	382	1,42
GERUND	340	1,26
Total annotated and unannotated words	Quantity	Quantity (%)
WORDS UNANNOTATED	20 223	42,84
WORDS ANNOTATED	26 983	57,16
Words annotated by the NOUN, ADJECTIVE, VERB categories	*21 845*	*80,96*
Words annotated by other categories	*5 138*	*19,04*

The Part-of-Speech Tagging Countercheck of unknown words was realized with the help of Levenshtein algorithm [4]. The algorithm revealed parts of speech of unknown words, picked up a possible correct form of the usage, and also gave an index of probability of correct forms. The stage of manual editing was carried out after computer-assisted Part-of-Speech detection: all parts of speech were checked by linguists-experts. In the case of the correct Part-of-Speech detection by the algorithm, this line of the table was marked as "true" (1). In an opposite case – "false" (0) (see Fig. 3).

ID	Seq -- specifies UNKNOWN words	DictSeq specifies words chosen by the algorithm	Similarity	PartOfSpeech	Ja&H
27073	рэфарміраванне	рэфармаванне	0.857142857	NOUN	1
27075	рэфарміраванню	фарміраванню	0.857142857	NOUN	1
27076	рэфарміравання	фарміравання	0.857142857	NOUN	1
27077	рэфектар	рэфлектар	0.888888889	NOUN	1
27078	рэфлэксаў	рэфлексаў	0.888888889	NOUN	1
27079	рэформ	рэформа	0.857142857	NOUN	1
27083	рэшаты	рэшата	0.833333333	NOUN	1
27086	рэштка	рэшткаў	0.857142857	NOUN	1
27089	рэшткай	рэштай	0.857142857	NOUN	1
27092	рэштку	рэшткі	0.833333333	NOUN	1
27094	рэюць	грэюць	0.833333333	VERB	1
27095	сілуэтам	сілуэтам	0.875	NOUN	1

Fig. 3. Countercheck of Annotated Categories in unknown words by the three linguists-experts

```
Dictionary contains 26983 entries

абмовіщца,VERB
адбірае,VERB
адзін,NUMERAL
адзінай,ADJECTIVE
адзінаццаць,NUMERAL
адзінкі,NUMERAL
адзінца,NOUN
адзіным,ADJECTIVE
адміністрацыйны,ADJECTIVE
азірнуўся,VERB
аналагічна,ADVERB
ані,PARTICLE
аніяк,ADVERB
архіттэктурнае,ADJECTIVE
архітэктурная,ADJECTIVE
архітэктурны,ADJECTIVE
асабіста,ADVERB
афіцыйнай,ADJECTIVE
афіцыйных,ADJECTIVE
аціраюцца,VERB
бабінцы,NOUN
большая,ADJECTIVE
вамі,PRONOUN
вашымі,PRONOUN
вельмі,ADVERB
відавочна,ADVERB
відавочныя,ADJECTIVE
```

Fig. 4. A fragment of the latest additional dictionary for the Belarusian *NooJ* module

The semi-automatic annotating of unknown words helped to form the version of the dictionary with annotated grammatical categories. It will be additional to the main dictionary, general_be.nod Dictionary of the Belarusian NooJ module (see Fig. 4). All parts of speech were tagged. The main difficulty was to find out the most effective way to generate all wordforms.

3 The Algorithm for Further Annotating of All Paradigms According to Inflection Classes in Nooj Format

The main concept is not only to get the category of a word but also the whole paradigm. The algorithm, which was worked out by the team, is the basis for the automatic generation of word paradigms. It consists of 16 consecutive interdependent steps. The algorithm outputs one or several most suitable paradigms of a word. It searches the nearest paradigm(s) in matches of the last letters of the word user needs to get the paradigm.

The algorithm for further annotating of all paradigms according to inflection classes is described below:

1. This step is used to search for a word in the dictionary of inflection classes. If the dictionary contains the word, then display to the user a complete paradigm and go to step 16. If a word is a homograph, then go to step 15. If the dictionary of inflection classes does not contain the word, then go to step 2.
2. This step is used to ask to the user to specify a part of speech of the word.
3. This step is depends on the part of speech chosen by the user to specify the grammatical features (with the possibility to leave the fill-in-the-blank fields empty if the user does not know the features).
4. This step is used to defines whether it is a changeable or an unchangeable word. If it is unchangeable, this steps displays a word with annotation, then goes to step 16. If the word is changeable, it goes to step 5.
5. This step is used to to take an unprocessed input word form for further processing, then goes to step 6.
6. This step is used to search in the dictionary for the words of specific inflection classes (with marked grammatical features) that end with the input word in current state. If the dictionary of inflection classes contains such words, then it goes to step 8, otherwise it goes to step 7.
7. This step is used to to remove the first letter of the input word in its actual state. Then it goes to step 6.
8. This step is used to divide the obtained words into "base" and "tail", where "tail" is a part of the obtained word matching with the input word in its actual state, and "base" ist the remaining part of the obtained word.
9. This step is used to select the "base" inside the original input word by cutting the "tail".
10. This step is used to separate the "tails" in other word forms of the obtained words, and to attach them all to the "bases" of the original input word.
11. If there more than one words are found, then it goes to steps 8–10 for all the words.

12. This step is used to compare the obtained paradigms. It deletes all the duplicates, leaving only unique paradigms.
13. If the user has given only one form, then it goes to step 14. If there are more unprocessed word forms given by the user as an input, then it goes to step 5 for the other word forms which the user has given, but it searches in the list of generated paradigms, not in the dictionary in step 6. If all word forms given by the user have been processed, then it goes to step 14.
14. If in the results only one paradigm is found, then it goes to step 15. If more than one paradigms are found, then it compares the obtained paradigms. It deletes all the duplicates, leaving only unique paradigms. Then it goes to step 15.
15. This step is used to give the word paradigms obtained by the user, and to ask to specify the correct one.
16. This step is used to end the algorithm.

Random 10 unknown words were selected for each of 14 categories to test the effectiveness of the algorithm. A total amount of words is 140 (see Fig. 5).

Short part-of-speech dictionary for Nooj			
word	**SK**	**category**	**COMMENTS**
абмакванне	,	NOUN	
болечка	,	NOUN	
вырай	,	NOUN	
нябытнасць	,	NOUN	
усіхнія	,	PRONOUN	
каторы	,	PRONOUN	
усенькае	,	PRONOUN	
чымсці	,	PRONOUN	
я	,	PRONOUN	
абароцісты	,	ADJECTIVE	

Fig. 5. A fragment of unknown words list for testing the algorithm

4   Word Paradigm Generator as the Software Prototype of the Algorithm

The software prototype of the algorithm, which is described above, is the Word Paradigm Generator service on the site www.Corpus.by (see Fig. 6) [5]. The user can specify multiple words of one paradigm by selecting a category with its grammatical attributes and clicking "Generate probable paradigms" button.

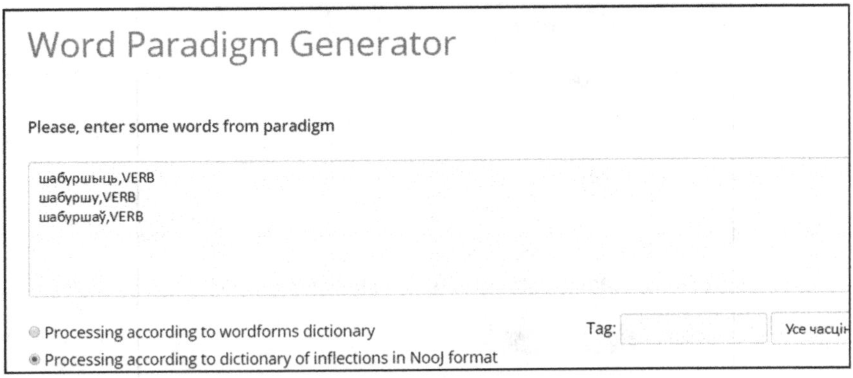

Word Paradigm Generator

Please, enter some words from paradigm

шабуршыць,VERB
шабуршу,VERB
шабуршаў,VERB

○ Processing according to wordforms dictionary
● Processing according to dictionary of inflections in NooJ format

Tag: [   ] Усе часці┤

Fig. 6. The interface of "Word Paradigm Generator"

The resource outputs several variants of the words of the same grammatical categories with inflection classes and their annotations. From the list of generated words, the user himself chooses the correct variant (see Fig. 7).

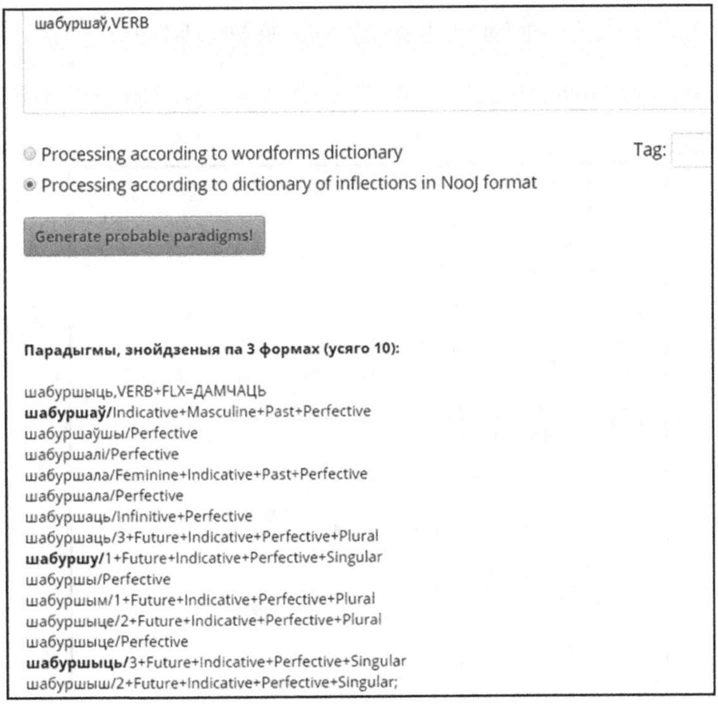

шабуршаў,VERB

○ Processing according to wordforms dictionary
● Processing according to dictionary of inflections in NooJ format

Tag: [   ]

Generate probable paradigms!

Парадыгмы, знойдзеныя па 3 формах (усяго 10):

шабуршыць,VERB+FLX=ДАМЧАЦЬ
шабуршаў/Indicative+Masculine+Past+Perfective
шабуршаўшы/Perfective
шабуршалі/Perfective
шабуршала/Feminine+Indicative+Past+Perfective
шабуршала/Perfective
шабуршаць/Infinitive+Perfective
шабуршаць/3+Future+Indicative+Perfective+Plural
шабуршу/1+Future+Indicative+Perfective+Singular
шабуршы/Perfective
шабуршым/1+Future+Indicative+Perfective+Plural
шабуршыце/2+Future+Indicative+Perfective+Plural
шабуршыце/Perfective
шабуршыць/3+Future+Indicative+Perfective+Singular
шабуршыш/2+Future+Indicative+Perfective+Singular;

Fig. 7. An output example of "Word Paradigm Generator" operation

It should be noted that only changeable parts of speech (noun, verb, adjective, participle, pronoun, and numeral) can be processed as they have a paradigm. The user can get annotation (tag) with stress arrangement of unchangeable parts of speech

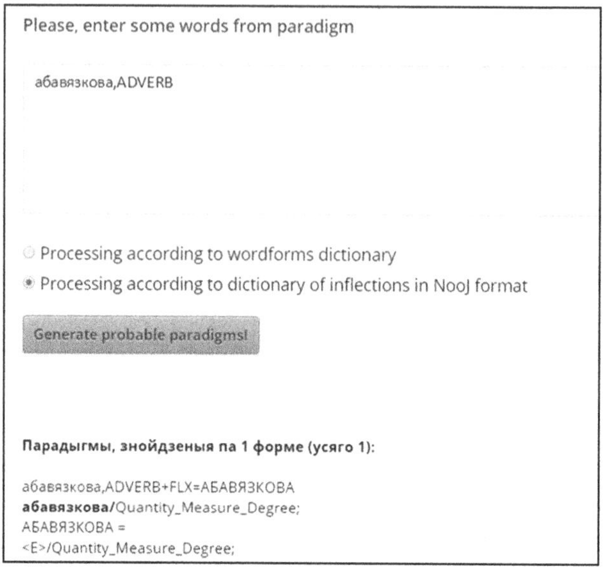

Fig. 8. Unchangeable adverb found in Nooj dictionary

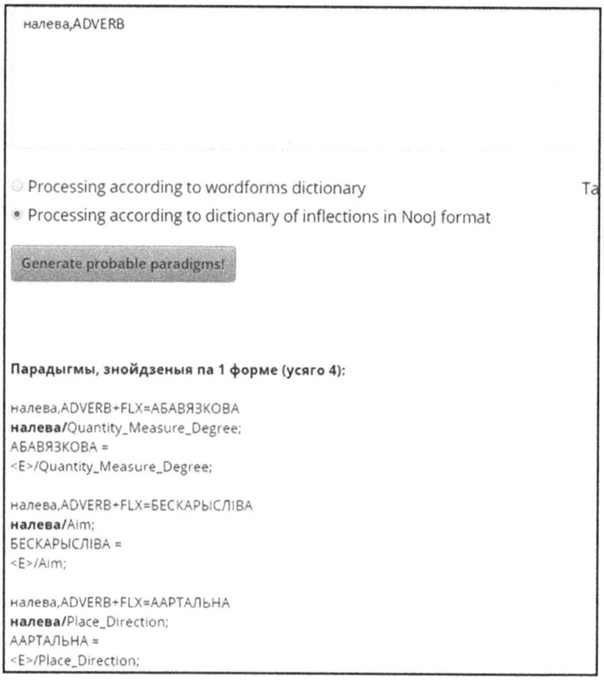

Fig. 9. Unchangeable unknown adverb processed by "Word Paradigm Generator"

Word Paradigm Generator

Please, enter some words from paradigm   ↺   X

```
клад,NOUN
кладзе,NOUN
кладамі,NOUN
```

○ Processing according to wordforms dictionary   Tag:
● Processing according to dictionary of inflections in NooJ format   Усе часціны мовы ▾

Generate probable paradigms!

Fig. 10. Searching for the paradigm of the word *"клад"* according to dictionary of inflections in NooJ format

(adverb, preposition, conjunction, particle, parenthesis, interjection, predicative, gerund) only if a word is found in the dictionary (see Fig. 8). Otherwise, he needs to choose the right variant among proposed. It would be better if the user could also indicate the tag of a word (see Fig. 9).

The most probable paradigm of the word "клад" (see Fig. 10), chosen by the expert after the word paradigm generation process in the service, is shown in Fig. 11.

Парадыгмы, знойдзеныя па 3 формах (усяго 11):

клад,NOUN+FLX=АВІЯСКЛАД
клад/Accusative+Common+Inanimate+Masculine
клад/Common+Inanimate+Masculine+Nominative
клада/Common+Genitive+Inanimate+Masculine
кладам/Common+Inanimate+Instrumental+Masculine
кладам/Common+Dative+Inanimate+Masculine+Plural
кладамі/Common+Inanimate+Instrumental+Masculine+Plura
кладах/Common+Inanimate+Masculine+Plural+Prepositional
кладзе/Common+Inanimate+Masculine+Prepositional
кладоў/Common+Genitive+Inanimate+Masculine+Plural
кладу/Common+Dative+Inanimate+Masculine
клады/Accusative+Common+Inanimate+Masculine+Plural
клады/Common+Inanimate+Masculine+Nominative+Plural;
АВІЯСКЛАД =
<E>/Accusative+Common+Inanimate+Masculine
+ <E>/Common+Inanimate+Masculine+Nominative
+ <E>a/Common+Genitive+Inanimate+Masculine
+ <E>ам/Common+Inanimate+Instrumental+Masculine
+ <E>ам/Common+Dative+Inanimate+Masculine+Plural

Fig. 11. A fragment of most probable paradigm of the word *"клад"*

5   Additional NooJ Dictionary (*general_be(add).dic*) on the Basis of Annotated Unknown Words

As a result, an additional NooJ dictionary (general_be(add).dic) for the Belarusian module was composed. A dictionary of 365 words was generated by "Word Paradigm Generator" in NooJ format. Every line provides the information about an unknown word, its part of speech, and a word from the dictionary "main, general_be.nod", which has the same paradigm. These two words of one line belong to the same inflection class (see Fig. 12).

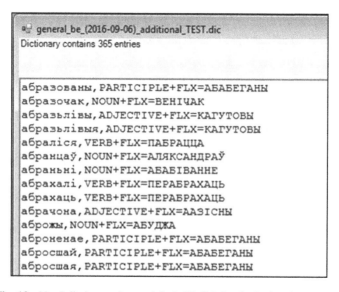

Fig. 12. NooJ dictionary (general_be(add).dic) for the Belarusian module

It takes approximately 0.035 of an hour (2.1 min) to process one word in "Word Paradigm Generator" by one linguists-experts. It means that we need nearly 945 h to annotate 27 thousand words. More detailed statistics are presented in Table 2.

Table 2. The Word Paradigm Generator level of efficiency according to its process by the user

Name	Quantity of words	Time consumed (h)
10 random words from 14 taken categories	140	4,9
First 365 words taken from an additional dictionary	365	12,775
An additional dictionary to be completely annotated	26 983	944,405

The corresponding table provides information about service efficiency. It should be noted that only changeable parts of speech due to NooJ system inflection class (+ FLX) [7], were tested. Unchangeable parts of speech were presented as follows: - (+ UNCH – "unchangeable").

6 Conclusion

Today, the algorithm has been created for annotating different categories and paradigms according to inflection classes were worked out. It was realized in the online prototype – "Word Paradigm Generator" (http://corpus.by/WordParadigmGenerator/). A list of unknown words extracted from Belarusian corpus was examined (365) and added to annotated words to the present dictionary on the basis of the Belarusian NooJ module. The next task is being planned: to develop the mechanism of automatic stress arrangement for all forms of an entire word on the basis of the Belarusian NooJ module.

References

1. Reentovich, I., Hetsevich, Y., Voronovich, V., Kachan, E., Kozlovskaya, H., Tretyak, A., Koshchanka, U.: The first one-million corpus for the Belarusian NooJ module. In: Okrut, T., Hetsevich, Y., Silberztein, M., Stanislavenka, H. (eds.) NooJ 2015. CCIS, vol. 607, pp. 3–15. Springer, Heidelberg (2016). doi:10.1007/978-3-319-42471-2_1
2. NooJ: A Linguistic Development Environment. Electronic resource (2015). http://www.NooJ4nlp.net
3. Hetsevich, Y.: Overview of Belarusian and Russian dictionaries and their adaptation for NooJ. In: Vučković, K., Bekavac, B., Silberztein, M. (eds.) Selected Papers from the NooJ 2011 International Conference on Formalising Natural Languages with NooJ, pp. 29–40. Cambridge Scholars Publishing, Newcastle (2011)
4. The Levenshtein-Algorithm. Electronic resource (2015). http://www.levenshtein.net/
5. Word Paradigm Generator. Electronic resource (2016). http://corpus.by/WordParadigmGenerator/
6. Hetsevich, Y.: Semi-automatic part-of-speech annotating for Belarusian dictionaries enrichment. In: Silberztein. M., Monteleone, M., Barone, L. (eds.) Proceedings of the NooJ 2016 International Conference (in print)
7. Silberztein, M.: Formalizing Natural Languages: The NooJ Approach. Wiley, London (2016)

Clinical Term Recognition: From Local to LOINC® Terminology. An Application for Italian Language

Francesca Parisi[(✉)]

Dipartimento di Ingegneria Informatica, Modellistica,
Elettronica e Sistemistica, Università della Calabria, Rende, Italy
francesca.parisi@dimes.unical.it

Abstract. This paper presents a methodology to recognize the specific domain terminology used by clinical and biological laboratories. The local linguistic forms have to be mapped with the standard LOINC® (https://loinc.org/) terminology and assigned to a specific identification code. The local linguistic forms recognition can give an important support for experts in finding the correct code and allows them the use of a local terminology not provided in the LOINC standard. The work presented a set of grammars build with the linguistic Nooj tool to recognize the local linguistic forms expressed in the Italian language. It shows how the linguistic elements contained in the clinical test definition can suggest important values of LOINC standard facilitating the coding process (e.g. the adjective "urinary" used in the local test suggests the values "urine" for the parameter "type of sample").

Keywords: Specific domain terminology · Semantic annotation · Automatic term recognition · Italian LOINC terminology · Nooj application

1 Introduction

The clinical data interoperability depends on the adoption of standard terminology allowing the information exchange between different institutions and maintaining the same meaning throughout clinical process. The use of these standards could be difficult because of the differences between local names and standard definitions.

All clinical laboratories have to codify their tests and results with LOINC international standard. Therefore, they have to map their local clinical terms and information to suitable LOINC concepts identifying the only one correct code. This mapping operation is sometimes difficult for clinical experts because they cannot find the correct code using their local linguistic forms.

A methodology to support linguistic interoperability of local laboratory terms and codes is presented in this paper with particular consideration for the Italian language terminology.

This paper is structured as follows. Section 2 presents the International Standard LOINC. Section 3 describes the context and the problems statement considered in

© Springer International Publishing AG 2016
L. Barone et al. (Eds.): NooJ 2016, CCIS 667, pp. 112–120, 2016.
DOI: 10.1007/978-3-319-55002-2_10

this paper. Section 3 illustrates the methodological approach with Nooj, Sects. 4 and 5 respectively present the preliminary results and conclusions.

2 The Standard LOINC

The international standard LOINC (Logical Observation Identifiers Names and Codes) is a universal coding system for identifying laboratory and clinical observations, facilitating the results exchange for clinical care, research, outcomes management, and many other purposes. Its general aim is to identify observations in electronic messages in Health Level Seven (HL7) format and observation messages, so that when hospitals, health maintenance organizations, pharmaceutical manufacturers, researchers, and public health departments receive such messages from multiple sources, they can automatically file the results in the right slots of their medical records, research, and/or public health systems [1].

Today, the standard LOINC is adopted widely and there are now more than 38,500 users in 167 different countries including Australia, Brazil, Canada, Cyprus, Estonia, France, Germany, Mexico, Mongolia, the Netherlands, Rwanda, Thailand, Turkey, and the United States. The LOINC content and associated documentation have been translated into many languages. In particular, the methodology presented in this paper is applied on the LOINC version translated into Italian [2]. LOINC codes promote the communication "machine to machine" allowing the exchange of clinical laboratory data between heterogeneous computing environments and the communication between experts identifying the exact nature of clinical tests.

The standard Loinc is composed of six different parts:

- the measured component or analyte name (e.g. Glucose);
- the observed property (e.g. substance concentration, mass, volume);
- the measurement timing (e.g. over time, momentary);
- the sample type or system (e.g. urine, serum);
- the measurement scale (e.g. quantitative, qualitative);
- the measurement method where relevant [3].

For each combination of these parts, a unique permanent code called "LOINC code" is assigned. This is the code that systems should use to identify test results in electronic reports. Each LOINC code corresponds to a particular data structure describing clinical tests characteristics with different detail levels that have to match perfectly with local analysis definitions. To support the LOINC coding process the Regenstrief Institute provides a Windows-based mapping utility called the Regenstrief LOINC Mapping Assistant (RELMA)® [4].

This tool facilitates the researches through the LOINC database and assists during the mapping operations between local and LOINC codes. It has specific search rules and works only with the official translated terminology and the synonyms recognized and stored in the LOINC database. For this reason, often the local linguistic forms used by clinical laboratories are not recognized in RELMA giving wrong results (or any results) during the codes research.

3 Methodological Approach with Nooj

The general aim of the methodology presented in this paper consists of recognizing specific linguistic elements such us adjectives, suffixes, prefixes and acronyms used by clinical laboratories. Others similar works such as [4, 5] concerning the automatic recognition of local linguistic clinical forms, but anyone work on the Italian Language.

The recognition of morphological and syntactic characteristics of local linguistic forms gives an important contribution to support the expert during the coding process. Such recognition allows experts to suggest the values of all others LOINC parameters (e.g.: starting from the definition, it is possible to deduce the value of "system" parameter; starting from "units of measure", it is possible to deduce the value of "property" parameter).

The linguistic and technical elements like specific adjectives, acronyms or the common affixes used in Italian Language make difficult the LOINC codes research with RELMA system because they are not associated to the official parts. As already explained, the official LOINC mapping system works only with the translated parts in different languages, and it cannot recognize the used local forms if they are not stored in the official db. Therefore, the local linguistic forms have to be associated to the specific LOINC parts if they are not recognized to Relma system (e.g.: the local form with suffix "emia" such "Glicemia" refers the analyte "Glucose"). This association "local forms – values of LOINC parameters" allows the experts to find the correct code in a relatively short lapse of time, and makes easy the coding process. For this reason, the methodology presented in this work could represent an important effort to support experts during the LOINC codes research and to improve the automatic recognition of local linguistic forms in the official mapping system RELMA. The linguistic elements considered in this paper to obtain some preliminary results are: the specific suffixes, the specific adjectives and the acronyms. The methodology has focused on the recognition of the linguistic elements contained in the clinical test definition that could suggest the values of others LOINC parts. In particular, "Table 1" and "Table 2" show an example of the suffixes and adjectives to suggest the value of the parameter "System" and Table 3 illustrates a subset of acronyms to identify the value of the parameter "Component".

Table 1. Suffixes for "System" parameter

Suffix	Type of sample/system
Emia	Sangue
Uria	Urine

3.1 The Corpus

The corpus considered in this work is composed of 5 clinical laboratory datasets. The methodology presented is carried out considering 488 local clinical test definitions. For identifying the most important linguistic variants, 30 typologies of clinical test are

Table 2. Adjectives for "System" parameter

Adjective	Type of sample/system
Urinario	Urine
Sierico	Siero
Plasmatico	Plasma

Table 3. Acronyms and symbols for "Component" parameter

Acronyms/symbols	Component (Italian version)
FSH	Follitropina
LH	Lutropina
P	Fosfato
K	Potassio
Na	Sodio
OH	Idrossido
HB	Emoglobina

considered. These generic typologies represent the most common exams carried out in clinical and biological laboratories (ex.: glycaemia, cholesterol, hemoglobin and so on).

3.2   The Nooj Dictionaries

The dictionaries built with Nooj tool describe the analytes (considering their acronyms) and the adjectives considered [7].

The acronyms are often used to indicate the elementary analyte (called the "core" of test definition) in combination with adjectives that could suggest the values of the LOINC parameters. Figures 1 and 2 show the analytes dictionary with their acronyms; Fig. 3 shows the dictionary of adjectives.

3.3   The Nooj Grammars

The general aim of the presented work is to create a linguistic resource to allow improving the LOINC codes research. Therefore, it needs to recognize all local linguistic forms and the information about LOINC parameters could be suggested. The syntactic grammars created with Nooj identify the morphological characteristics of clinical definitions (e.g.: the suffixes "emia" and "uria") and the linguistic parts (e.g.: adjectives "urinario", "sierico", "plasmatico", "ionizzato") to deduce others important values for LOINC parameters. This recognition is an important support during the coding process; it allows taking the correct decision about codes assignment [8]. In particular, the suffixes and the adjectives that could suggest the value of the "System" parameter, how illustrated in the grammars showed in Figs. 4 and 5.

```
Azoto,N+Analita+FLX=F1
Calcio,N+Analita+FLX=F1
Glucosio,N+Analita+FLX=F3
Albumina,N+Analita+FLX=F2
Sodio,N+Analita+FLX=F1
Potassio,N+Analita+FLX=F1
Cortisolo,N+Analita+FLX=F1
Lipidi,N+Analita+FLX=F5
Bilirubina,N+Analita+FLX=F2
```

Fig. 1. Extract of analytes Nooj dictionary/analytes

```
G,Glucosio,N+Analita+Acronimo
Na,Sodio,N+Analita+Acronimo
K,Potassio,N+Analita+Acronimo
FSH,Follitropina,N+Analita+Acronimo
LH,Lutropina,N+Analita+Acronimo
P,Fosfato,N+Analita+Acronimo
OH,Idrossido,N+Analita+Acronimo
Hb,Emoglobina,N+Analita+Acronimo
HBA1,Emoglobina A1,N+Analita+Acronimo
HBA2,Emoglobina A2,N+Analita+Acronimo
HBF,Emoglobina Fetale,N+Analita+Acronimo
HGB,Emoglobina A1c,N+Analita+Acronimo
HbA1c,Emoglobina A1c,N+Analita+Acronimo
GPT,Aspartato transaminasi,N+Analita+Acronimo
AST,Aspartato transaminasi,N+Analita+Acronimo
GPT,Alalina aminotransferasi,N+Analita+Acronimo
ALT,Alalina aminotransferasi,N+Analita+Acronimo
LDH,Lattato deidrogenasi,N+Analita+Acronimo
```

Fig. 2. Extract of analytes Nooj dictionary/acronyms

```
Urinario,A+AgLoincUrine+FLX=AG1
Sierico,A+AgLoincSiero+FLX=AG3
Ionizzato,A+AgLoincSiero+FLX=AG2
Plasmatico,A+AgLoincPlasma+FLX=AG3
```

Fig. 3. The LOINC adjectives Nooj dictionary

The combination between Analytes dictionary elements and specific adjectives (e.g.: "urinario", "sierico", "ionizzato", "plasmatico") are annotated with the information about the suggested values for the "System" LOINC parameter (ex.: the adjective "urinario" suggests the value "urine", the adjective "sierico" or "ionizzato suggests the value "siero" and the adjective "plasmatico" suggests the value "plasma"). The notion "<N+Analita>" allows recognizing also the inflectional forms "<N+Analita+ Acronimo>".

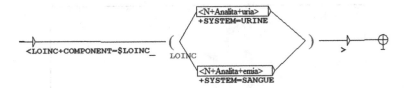

Fig. 4. The grammar for the suffix "emia" and "uria"

Fig. 5. The grammar for LOINC adjectives

4 Preliminary Results

In this section, we present the results obtained with the linguistic analysis carried out with Nooj. In the Figs. 6 and 7, the results concerning semantic annotation of the linguistic forms with the suffixes "emia and "uria" are presented. The syntactic grammar showed in the Fig. 4, allows creating the xml elements called "LOINC" with the attributes "Component" and "System". These attributes represent the suggested values for the Loinc parameters.

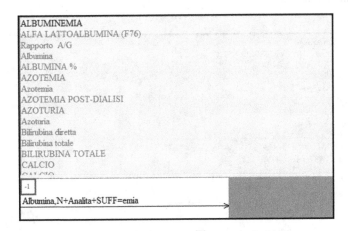

Fig. 6. Semantic annotation: suffix "emia"

For the term "Albuminemia" (Fig. 6), Nooj allows knowing that it derives from the lexical unit "Albumina", so it is an inflectional forms that takes the suffix "emia" (e.g.: *"Albumina,N+Analita+SUFF=emia"*). The Figs. 8 and 9 show the semantic

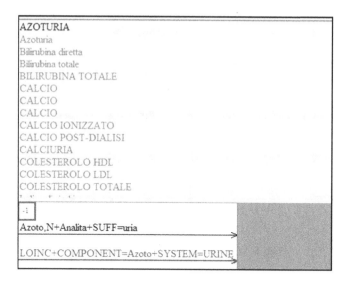

Fig. 7. Semantic annotation: suffix "uria"

annotation for the forms "Analyte+Adjective". The first one shows how an acronym is an inflectional form for the lexical units contained in the Analytes dictionary and the value suggested by the adjective (e.g.: "FSH" is an acronym for the lexical unit "Follitropina" and the adjective "urinario" suggests the value "urine" for the "System" parameter). The second one shows the annotation for the same syntactic structure with the lexical unit "Calcio" and the adjective "ionizzato". It suggests the value "siero" for the "System" parameter.

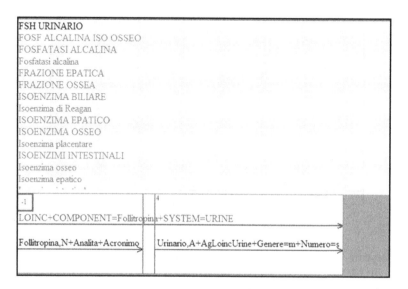

Fig. 8. Semantic annotation: Acronym + Adjective (suggested value: "urine")

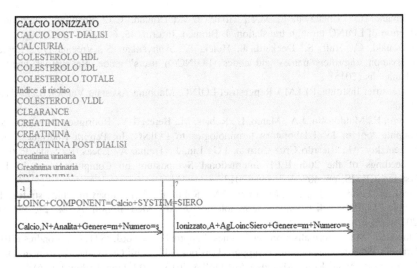

| CALCIO IONIZZATO |
| CALCIO POST-DIALISI |
| CALCIURIA |
| COLESTEROLO HDL |
| COLESTEROLO LDL |
| COLESTEROLO TOTALE |
| Indice di rischio |
| COLESTEROLO VLDL |
| CLEARANCE |
| CREATININA |
| CREATININA |
| CREATININA POST DIALISI |
| creatinina urinaria |
| Creatinina urinaria |

Fig. 9. Semantic annotation: Analyte + Adjective (suggested value: "siero")

5   Conclusion and Perspectives

In this paper, a work in progress carried out with the Nooj linguistic tool was presented. The methodology proposed could be an important starting point for the construction of an automatic recognition of clinical and biological test definition forms that have to be mapped to official LOINC terminology.

The paper has outlined the importance of coding operations but also the problems encountered by the clinical expert during this process. It is evident the importance of domain specific human knowledge during coding process of clinical data for making the correct association with official codes.

Accordingly, the methodology presented aims to describe the knowledge contained in specific domain terminology and supporting the experts during the association with healthy standard terminology and informatics structure. The methodology considered only a subset of part of speech elements that could help the coding process.

The future goal aims to recognize all elements that could suggest the values of LOINC parameters and create a linguistic resource to give an important effort for improving the LOINC code research starting to the clinical test definition and using also the local forms. The units of measurement will be annotated to deduce the "Property" and "Timing" LOINC.

References

1. McDonald, C.J., Huff, S.M., Suico, J.G., Hill, G., Leavelle, D., Aller, R., Williams, W.: LOINC a universal standard for identifying laboratory observations: a 5-year update. Clin. Chem. **49**, 624–633 (2003)

2. Vreeman, D.J., Chiaravalloti, M.T., Hook, J., McDonald, C.J.: Enabling international adoption of LOINC through translation. J. Biomed. Inform. **45**, 667–673 (2012)
3. McDonald, C., Huff, S., Deckard, J., Holck, K., Abhyankar, S., Vreema D.J.: Logical observation identifiers names and codes (LOINC®) users' guide. Regenstrief Institute, Indianapolis (2015)
4. Rengenstrief Institute, RELMA Rengestrief LOINC, Mapping Assistant Version 6.12, Users' Manual
5. Parcero, E., Maldonado, J.A., Marco, L., Robles, M., Berez, T.V., Rodriguez, M.: Automatic mapping tool of local laboratory terminologies to LOINC. In: Pereira Rodrigues, P.A., Pechenizkiy, M., Ricardo Cruz Correia, J.G., Liu, J., Traina, A., Lucas, P., Soda, P. (eds.) Proceedings of the 26th IEEE International Symposium on Computer-Based Medical Systems, CBMS, pp. 409–412, June 2013
6. Lee, L.H., Groß, A., Hartung, M., Liou, D.M., Rahm, E.: A multi-part matching strategy for mapping LOINC with laboratory terminologies. J. Am. Med. Inform. Assoc. **21**, 792–800 (2014)
7. Silberztein, M.: La formalisation des langues: l'approche de NooJ. ISTE Ed.: Londres (2015)
8. Parisi, F.: Semantic annotation to support decision-making. In: Gudivada, V., Roman, D., di Buono, M.P., Monteleone, M. (eds.) ALLDATA 2016, p. 99. Iaria, Lisbon (2016)

Enumerative Series in Spanish: Formalization and Automatic Detection

Walter Koza[✉]

Instituto de Literatura Y Ciencias Del Lenguaje,
Pontificia Universidad Católica de Valparaíso, Proyecto FONDECyT,
11130469 Valparaíso, Chile
walter.koza@pucv.cl

Abstract. Analysis and formalization are presented about the enumerative series structure in Spanish for subsequent computational implantation. The enumerative series is a textual construct composed by a matrix, an enumerator and an enumeration. Each element of an enumeration is called 'enumerating'; all elements are related to an "enumeratheme". Enumeratings and enumeratheme establish a kind of hypernym-hyponym relationship (e.g. in "the days Monday, Tuesday, and Wednesday"; "days" is the enumeratheme and "Monday", "Tuesday", and "Wednesday", the enumeratings). According to this description, a formalization is achieved allowing computational implantation for the detection of enumerations and enumerative series. To accomplish this objective, the NooJ program is used, and the methodology is tested on a corpus of Wikipedia entries related to the medical field, reaching 100% precision, 52.50% recall, and 68.65% F measure.

Keywords: Enumerative series · Enumeration · Enumerating · Enumerator · Automatic analysis

1 Introduction

A study of the enumerative series in a sentence from a formal perspective is proposed. The enumerative series contains three elements:

- An enumeration: an ordered set of "enumeratings"
- An enumerator: an element syntactically related to the enumeration
- An enumeratheme: the hypernym of each enumerating (it can be explicit or not) [1].

For example:
Juan trabaja los días lunes, martes, miércoles y sábados.
[John works on Mondays, Tuesdays, Wednesdays and Saturdays.]
In this sentence, the three elements mentioned below are as follows:

1. Enumeratings: 'lunes' ['Mondays'], 'martes' ['Tuesdays'], 'miércoles' ['Wednesdays'], and 'sábado' ['Saturdays'].
2. Enumerator: 'trabaja' ['works'].
3. Enumeratheme: 'días' ['days'].

© Springer International Publishing AG 2016
L. Barone et al. (Eds.): NooJ 2016, CCIS 667, pp. 121–131, 2016.
DOI: 10.1007/978-3-319-55002-2_11

Following this criteria, an analysis is proposed which contemplates the nature of the enumeratings, their syntactic relationships and the enumerator. According to this analysis, a computational implantation that detects enumerations and enumerative series in natural language texts is made. This methodology is tested in a corpus composed of medical domain Wikipedia entries, which added up to a total of 57632 words and reached 100% precision, 52.50% recall, and 68,65% F measure.

The article is organized as follows: in Sect. 2, an inquiry about the enumerative series is made. In Sect. 3, a description of the computational implantation is given. In Sect. 4, there is an analysis of the results obtained. Finally, in Sect. 5, conclusions derived from our research are presented.

2   About the Enumerative Series

The enumerative series is a textual structure that has been studied from diverse perspectives, such as rhetoric [2], discourse analysis [3–5], textual grammar [6], analysis of oral discourse [7], and from a computational linguistics [8, 9]. In the case of Spanish, the most exhaustive work has been *La serie enumerativa en el discurso oral en español*, by Cortés [10]. This volume compiles works focused on the enumerative series phenomenon in oral texts and from the perspectives of rhetoric, semantics and pragmatics.

Cortes [10] defines the enumerative series as a set of enumerated elements (enumeratings) which develops the thematic progression of a matrix or 'enumeratheme'. The enumeration interpretation consists in identifying it with an enumeratheme; and, in the case that it is not explicit, the enumeratheme can be inferred starting from the enumeratings [1].

The present work focuses on the syntactic relationship between the enumeration and the rest of the sentence. For that objective, the enumerator category is proposed, in order to refer to the element which triggers the enumeration and is syntactically related to each enumerating. The enumerator can coincide or not with the enumeratheme.

Juan visitó parques, museos y restoranes.
[John visited parks, museums and restaurants.]

Here, the function of enumerator is expressed by the verb 'visitó' ['visited'] and the enumeratheme ('lugares' ['places']) is not explicit. Also, in some cases, the enumerators are enumeratings of a previous enumeration:

Juan visitó parques, museos y restoranes argentinos, peruanos y mexicanos.
[John visited parks, museums and Argentinian, Peruvian and Mexican restaurants.]

The last element of the first enumeration, 'restoranes' ['restaurants'], is the first enumerator that initiates the second.

When the enumeratheme and enumerating coincide, there is a relationship of apposition between the enumeration and the enumerator, as seen in the next example.

Me gusta lo típico; viajar, bañarme, la lectura, el arte.
[I like typical things; traveling, taking a shower, reading, art.] (Taken from [11]).

Graphically, this structure can be structured as follows:

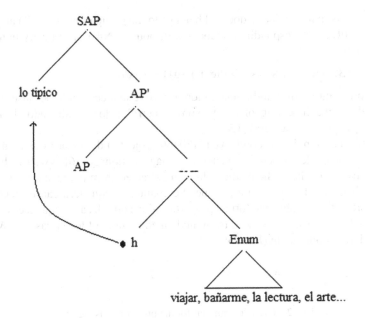

Fig. 1. Structure of the phrasal apposition [12], with an enumeration as an appositive term.

In Fig. 1, it can be observed how the enumeratheme 'lo típico' ['typical'] forms and apposition with the enumeration in its totality, at the same time establishing a hypernym-hyponym relationship with each enumerating.

3   Computational Implantation

A methodology for the automatic recognition of enumerations and enumerative series in natural language texts is developed. This procedure is carried out with the objective of obtaining a tool that could be applied to diverse tasks, such as the automatic extraction of term candidates [13]. To that purpose, the free software NooJ is applied [14]. This program has a variety of tools for the natural language treatment:

- Morphologic and derivational grammar (.nof files): inflection and derivation models.
- Dictionaries (.dic files): lists of words with diverse linguistic information.
- Productive grammar (.nom files): regular systems or graphics useful for the treatment of chains of characters with determined formal properties.
- Syntactic grammar (.nog files): regular systems or graphics useful for the treatment of chains of characters formed by two or more lexical units, generally separated by blank spaces.

Morphological grammar allows the generating of words variations. For instance, a word like 'médico' ['doctor'] is listed as follows:

médico, N + FLX = N4

That means that 'médico' ['doctor'] belongs to the group of nouns ('N') and has an inflection ('FLX') corresponding to model 4 for nouns ('N4'), specifically in morpho-logic grammar:

N4 = (<B>o/masc + sg | <B>a/fem + sg) | <B>s/pl;

This procedure proves to be much more effective and economic because the same model allows the inflecting of many words ('perro' ['dog'], 'abogado' ['lawyer'], 'sobrino' ['nephew'], etcetera) [15].

On this occasion, in the Fondecyt 11130469 project, a dictionary was compiled in NooJ comprised of lemmas from a general language dictionary [16] as well as the nouns and adjectives specific to the medical domain extracted from the medical dictionaries [17, 18]. With productive grammar, specific character sequences can be recognized through previously established formal properties. For example, any sequence comprised by a capital letter (<U>) followed by an undefined number of lower cases (<W>) can be tagged as a proper noun (Fig. 2):

Fig. 2. Graphic grammar for the proper nouns tagging

Syntactic grammars can operate in an isolated way, or they may interact with productive grammars and dictionaries, which, in turn, are integrated with inflective, derivative and productive grammar. In the first case, the grammar entries would be words, for example, DP = 'el' ['the'] + 'médico' ['doctor']. Another option could be categories declared in the dictionary. In this way, DP = DET +N, this grammar will recognize phrase every determining sequence and noun as a determining as long as they have been previously recognized by a dictionary or by productive grammar.

In the following section, we describe the computational implantation.

3.1   Lexical Analysis and Recognition of Punctuation Marks

Although the NooJ dictionary is vast, it is impossible to update, constantly and manually, all dictionaries and terminological bases. Thus, to mitigate the risk of words that have not been analyzed, productive grammars are established. One of these productive gram-mars is labeled as a "proper noun" – which means: any sequence of letters starting with a capital letter. The problem is that every word that initiates a sentence is also labeled as "proper noun". For example, in the fragment of the corpus: "Artículo principal: (…)", in a preliminary tagging, NooJ generated the following result:

To avoid this conflict, the program allows to organize dictionaries and grammars hierarchically. To this purpose, it is specified in the program preferences, that the dictionary 'Diccionario Enumeración.nod' has a higher priority (H1), while the produc-tive grammar of the proper nouns 'NPR.nom' possesses a lower priority (L1) (Fig. 3):

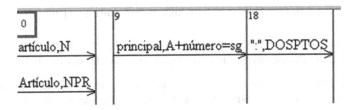

Fig. 3. Fragment of the NooJ lexical analysis

In this way, the new analysis result is as follows (Fig. 4):

Priority	Resource
H1	Diccionario Enumeracion.nod
L1	NPR.nom

Fig. 4. Lexical hierarchy

The results proceeded in a similar way with morphologic clues such as: every word ending in '-ción' is a singular feminine noun, or every word ending in '-ó' is an indicative verb (Fig. 5).

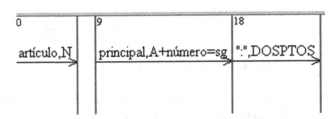

Fig. 5. New output of NooJ

After the lexical analysis, a new syntactic grammar is developed for the recognition of enumerative series.

3.2 Syntactic Grammar: Phrase Recognition, Enumerations and Enumerative Series

Syntactic grammars are established for the recognition of lexical phrases (nominal, adjectival, verbal and prepositional). In order to diminish the risk of ambiguities, the analysis started with the formation of nucleus phrases [19]. The nucleus phrases are phrases shaped by fixed morphosyntactic categories. Such categories allow to establish where these phrases start, end, how they are shaped and which is their nucleus. Based on given properties of linearity, combinatorial possibilities are restricted from their elements:

- Nucleus Nominal phrase (nnp): 'los más bellos paisajes' ['the most beautiful sceneries']
- Nucleus Adjective Phrase (nadjp): 'más importantes' ['more important']
- Nucleus Verbal Phrase (nvp): 'no los compraron' ['the did not buy it']

Abney [19] justifies the analysis in nucleus phrases because they allow an automatic syntactic analysis with fewer difficulties. In the case of the examples, it is possible to determine when 'los' ['the' (plural)] corresponds to a neutral article (nnp) and when it corresponds with a clitic pronoun (nvp).

Once the nucleus phrases are determined, the formation of complete phrases proceeded. In this occasion, the subordinate clauses are not taken into consideration nor in the case of the verbal phrases (VP), the specifier DP. As an example, the structure of the NP is presented:

- NNP + NADJP + [(PREP + NNP) ≥ 1] = NP
- NNP + NADJP = NP
- NNP = NP

Graphically (Fig. 6):

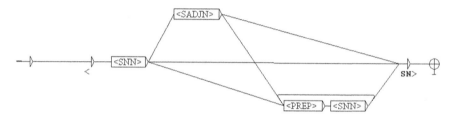

Fig. 6. Grammar for the NP recognition

Once the phrases are established, the formation of grammar for the recognition of enumerations proceeded. In this way, if, for example, the clause (2) ('Juan trabaja los días lunes, martes, miércoles y sábados' ['John works on Mondays, Tuesdays, Wednesdays and Saturdays']) is taken, a pertinent model for the enumeration would be:

NP + NP + comma + NP + coor. + NP = NPENUMERATION

This means to say that, if in the text, an NP followed by a comma is found, followed by another NP followed by a coordinating conjunction, plus another NP; then there is a NP enumeration.

In this manner, a group of rules that model the possible cases of phrasal enumerations is developed:

- (XP + comma) ≥ 1 + XP + conjunction + XP = XPENUMERATION
- (XP + comma) ≥ 3 + ellipsis = XPENUMERATION
- (XP + comma) ≥ 3 + 'etcetera' = XPENUMERATION
- (XP + comma) ≥ 3 = XPENUMERATION (in cases of asyndeton)

These structures are embedded computationally through the following grammar (Fig. 7):

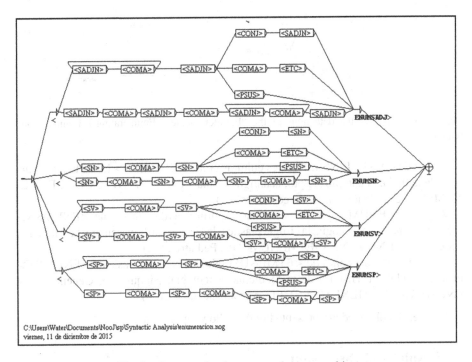

C:\Users\Water\Documents\NooJ\sp\Syntactic Analysis\enumeracion.nog
viernes, 11 de diciembre de 2015

Fig. 7. Grammar for the enumerations recognition

In this way, recognition is achieved of the mentioned phrasal enumerations. For example, let use exasmine some fragments of the NooJ output:

(…) estas sustancias eran utilizadas en rituales mágicos por <ENUMSN> chamanes, sacerdotes, magos, brujos, animistas, espiritualistas o adivinos </ENUMSN>. (…)

(…) valor <ENUMSADJ> legal, educacional, informativo y científico </ENUMSADJ>, (…)

(…) los trastornos del crecimiento <ENUMSP> de las células, de los tejidos y de los órganos </ENUMSP> (…)

(…) suficientemente peligrosa como para <ENUMSV> retrasar, modificar o contraindicar la operación </ENUMSV> (…)

Finally, a syntactic grammar is developed for the detection of the "Enumeratheme - Enumerating - Enumeration" sequence. Expressions like 'tales como' ['such as'], 'como por ejemplo' ['for example'], 'son' ['they are'], 'los que incluyen' ['including'] and 'los cuales incluyen' ['which includes']. Colons and opening parenthesis punctuation marks are also included when the explicit enumeration connector was not available (Fig. 8).

Thus, it is possible to detect structure like the following:

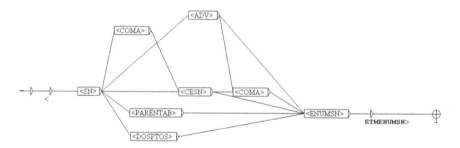

Fig. 8. Grammar for the detections of the "Enumeratheme - Enumerating - Enumeration" sequence

(…) <ETMENUMSN> diferentes culturas como <ENUMSN> la medicina Ayurveda de la India, el antiguo Egipto, la antigua China y Grecia </ENUMSN> </ETMENUMSN>. Uno de los primeros recorridos (…)

(…) <ETMENUMSN> las especialidades quirúrgicas de la medicina: <ENUMSN> la cirugía general, la urología, la cirugía plástica, la cirugía cardiovascular y la ortopedia, entre otros </ENUMSN> </ETMENUMSN>. Pediatría, (…)

(…) de salud, así como de <ETMENUMSN> las ciencias básicas (<ENUMSN> Física, Estadística, Historia de la Medicina, Psicología, Bioquímica, Genética… </ENUMSN> </ETMENUMSN>). El tercer año se dedica (…)

The results obtained are presented in the following section.

4   Results and Discussion

We have been testing the method here presented on a corpus of 64 Wikipedia entries pertaining to the medical field. The total corpus contained 57632 words and, after a manual revision, 357 enumerations are recognized which are distributed in the following manner: Nominal: 319 (105 of these with an explicit enumeratheme); Adjectival: 26; Verbal: 7; and Prepositional: 5.

Once the automatic analysis is done, 370 expressions tagged as enumerations are extracted, presenting 322 of which are correctly recognized; 42 not recognized and 32 erroneously tagged. This implied 88,46% precision, 90,20% recall, and 89,82% F measure. The results for each type of enumeration are divided as follows (Table 1):

Table 1. Results obtained.

Enumeration	Total in the Corpus	Recognized	Not Recognized	Wrong Tagged	Precision	Recall	Measure F
Nominal	319	288	31	36	88.88%	90.28%	89.57%
Adjectival	26	23	3	4	85.18%	88.46%	86.79%
Verbal	7	7	0	2	71.43%	100%	55%
Prepositional	5	4	1	0	100%	83%	91%

Lexical ambiguities are the main problem in the automatic recognition, for example:

(…) formación sanitaria especializada para <ENUMSADJ> Médicos, Farmacéuticos y otros graduados </ENUMSADJ>/ licenciados universitarios (…)

Here, the enumeration is tagged as an adjective because 'médico' ['doctor'] and 'farmacéutico' ['pharmaceutical'] are in the dictionary with noun and adjective tags. On the other hand, 'graduado' ['graduate'] appears as an adjective and verb. Contrastingly, there is also evidence of problems with proper nouns. For example:

(…) personajes tales como (…) William Coley, James D. Watson, Salvador Luria, Alexandre Yersin, Kitasato Shibasaburo, Jean-Martin Charcot, Luis Pasteur, Claude Bernard, Paul Broca, Nikolái Korotkov, William Osler y Harvey Cushing como los más importantes entre otros. (…)

As previously mentioned, a productive grammar was established which recognized all sequences starting with a capital letter followed by one or more lowercase letters as a proper noun. However, we may observe that the enumerating "James D. Watson" contains only the initial of the second name, followed by a period, and "Jean-Martin Charcot" contains a hyphen. These particular cases completely prevented the enumeration to be detected.

Furthermore, on the syntactic plane, there is evidence of two types of errors: appositions formed by a nominal phrase followed by another compound (with two coordinated nucleus) and cases in which an adjunct verb (generally of place or time) appears displaced to the left, followed by NP coordination.

(…) <ENUMSN> Dos investigadores, Roger Guillemin y Andrew Schally </ENUMSN>, observaron (…)

(…) en <ENUMSN> un hospital de Edimburgo, el tocólogo James Simpson y su compañero Dunkan </ENUMSN> practicaron el primer parto sin dolor (…)

In the case of verbal enumerations, only the non-finite verbs (participles, infinitives, and gerunds) are considered. The two erroneously tagged correspond to an overlap of tags:

(…) no filiada mediante <ENUMSN> el <ENUMSV> legrado, cepillado y lavado </ENUMSV> bronquial </ENUMSN>, con aspiración citológica (…)

(…) analizadores clínicos <ENUMSV> <ENUMSADJ> automatizados, computarizados y especializados </ENUMSADJ> en diferentes campos analíticos </ENUMSV> como hematología (…)

Regarding prepositional enumerations, only five are counted, out of which 4 are successfully recognized. The fifth cannot be analyzed due to the complexity of one of the NP that acted as a complement of the preposition:

(…) distintas técnicas: de osteosíntesis, de traslado de tejidos mediante colgajos y trasplantes autólogos de partes del cuerpo sanas a las afectadas, etc. (…)

Concerning the "Enumeratheme - Enumerating - Enumeration" sequence, that is to say, the enumerative series, 50 structures of this type are detected, and there are no erroneous detections, which implies 100% precision, 52.50% recall, and 68.65% F measure. Although the results are, to a certain degree, positive, this aspect of the enumeration is intended to be developed in the next stages of work.

5   Conclusion

This study of the enumerative series is carried out with the objective of establishing a model of automatic recognition in natural language texts. For this purpose, some issues concerning this enumeration series construction are analyzed. The methodology is evaluated in a corpus composed of Wikipedia entries in the medical field, and adequate percentages are achieved in precision, recall, and F measure. However, it is possible to observe a series of inconveniences in the detection, which is why the refining of some rules is expected, in order to improve results.

Future work will be organized according to the following steps: (i) enumerations modeling and amplification of the automatic detection method, including enumerations of greater complexity (enumeratings separated by semicolons, bullet points, etcetera); (ii) emphasis on the analysis of the enumeration within the syntax-discourse interface studies; and (iii) establishing a connection with the research of Fauconnier, Kamel and Rothenburger [9], structuring experiments for the automatic recognition of enumera-theme-enumeration in Spanish.

References

1. Ho-Dac, L., Woodley, M., Tanguy, L.: Anatomie des structures énumératives (2010). http://www.iro.umontreal.ca/~felipe/TALN2010/Xml/Papers/all/taln2010_submission_26.pdf
2. Marchese, A., Forradellas, J.: Diccionario de Retórica. Crítica y Terminología Literaria. Ariel, Barcelona (1986)
3. Bras, M., Prévot, L., Vergez-Couret, M.: Quelle(s) relation(s) de discours pour les structures énumératives? In: Durand, J., Habert, B., Laks, B. (eds.) Actes du Colloque Mondial de Linguistique Française CMLF 2008, pp. 1945–1964. Institut de Linguistique Française, Paris (2008)
4. López Samaniego, A.: Los ordenadores del discurso enumerativo en la sentencia judicial: ¿Estrategia u obstáculo? Revista de Llengua i Dret **45**, 61–87 (2006)
5. Vergez-Couret, M., Bras, M., Prevot, L., Vieu, L., Attalah, C.: Discourse contribution of Enumerative structures involving pour deux raisons. In: Proceedings of Constraints in Discourse (2011) http://www.irit.fr/publis/LILAC/VCBPVA-pour2raisons-CID11.pdf
6. Porhiel, S.: Les structures énumératives à deux temps. Rev. Romane **42**, 103–135 (2007)
7. Cortés, L.: La serie enumerativa en el cierre de los discursos. Estud. Filológicos **49**, 39–57 (2012)
8. Luc, C.: Une typologie des structures énumératives basée sur les structures rhétoriques et archictecturales du texte. In: TALN (eds.) Actes de TALN 2001, pp. 263–272. TALN, Tours (2001)
9. Fauconnier, J., Mouna, K., Rothenburger, B.: Une typologie multi-dimensionnelle des structures énumératives pour l'identification des relations termino-ontologiques. In: Aguado de Cea, G., Aussenac-Gilles, N. (eds.) Conférence Internationale sur la Terminologie et l'Intelligence Artificielle (TIA 2013) pp. 137–144, Université Paris 13, Paris (2013)
10. Cortés, L.: La serie enumerativa en el discurso oral en español. Arco Libros, Madrid (2008)
11. Camacho, M.: Relaciones textuales entre serie y matriz. In: Cortés (coord.) La serie enumerativa en el discurso oral en español, pp. 127–155, Arco Libros, Madrid (2008)

12. Moro, A.: Dynamic antisymmetry. Movement as a symmetry breaking phenomenon. Studia Lingüística **51**, 50–76 (1997)
13. Koza, W.: Propuesta de extracción automatica de candidatos a término del dominio medico procesando información lingüística. Descripción y evaluación de resultados. Alfa. Revista de Linguistica **59**(1), 113–127 (2015)
14. Silberztein, M.: Formalizing Natural Languages. The NooJ Approach. Wiley, London (2016)
15. Bonino, R.: Una propuesta para el tratamiento de los enclíticos en NooJ. Infosur Revista **7**, 31–40 (2015)
16. RAE: Nueva Gramática de la Lengua Española, Asociación de Academias de la Lengua Española, Buenos Aires (2006)
17. Diccionario Mosby (versión electrónica) Harcourt, Madrid (2005)
18. Cárdenas, E.: Terminología Médica. Mc Graw Hill, México (2012)
19. Abney, S.P.: Parsing by Chunks. In: Berwick, R.C., Abney, S.P., Tenny, C. (eds.) Principle-Based Parsing, pp. 257–278. Springer Science + Business Media, Dordrecht (1994)

Recognition and Extraction of Latin Names of Plants for Matching Common Plant Named Entities

Mohamed Aly Fall Seideh[1(✉)], Hela Fehri[1], and Kais Haddar[2]

[1] MIRACL Laboratory, Higher Management Institute of Gabes,
University of Gabes, Gabès, Tunisia
almedyfall@gmail.com, helafehri@yahoo.fr
[2] MIRACL Laboratory, Faculty of Science of Sfax, MIRACL,
University of Sfax, Sfax, Tunisia
kais.haddar@yahoo.fr

Abstract. The aim of this work is to use the Latin names of plants to make the pairing of French-Arabic plants common names. The scientific name of a plant is a common denominator between the common names because the scientific name is in Latin and is independent of other languages. To do the pairing, we will firstly identify and extract the Latin names of plants by using the NooJ platform, then from the sets: common names of plants of the French corpus (S1), the Latin names of the French corpus (S2), common names of plants of the Arabic corpus (S3) and Latin names of the Arabic corpus (S4), we model, implement and test a matching approach.

Keywords: Named entity recognition · Matching · Nooj

1 Introduction

Common plant names often seem easier to remember than scientific names, but they are not as precise. Not only a common name refers to very different plants, conversely a single species can have more than one common name. This can lead to confusion. While it is quite appropriate to use common names when everyone knows what you mean, often it's much better to use the proper botanical name. A botanical name or Latin name or scientific name consists of two words, and is therefore referred to as a "binomial". The first word represents the larger group the plant belongs to, the genus, and its first letter is always capitalized. The second word is the species and is always lowercase. There cannot be more than one identical species name in each genus. The species name is often descriptive of some aspect of the plant. The genus name can be used alone when discussing a group of plants, but the specific epithet is never used by itself.

The main objective of this work is to make the matching between the named entities (NEs) extracted from French Arabic parallel corpora. The matching is the action to link each French identified named entity to its Arabic correspondence, thus a lexicon will be generated. The matching will be based on the Latin names of plants. The next section will explain in detail the Problematic and resolution process.

© Springer International Publishing AG 2016
L. Barone et al. (Eds.): NooJ 2016, CCIS 667, pp. 132–144, 2016.
DOI: 10.1007/978-3-319-55002-2_12

Section 3 is devoted to the study of Latin plants names and will focus on the three most used components: family name, genus and epithet.

The Sect. 4 copes with the proposed method for the recognition and extraction of Latin plants names and the matching Latin names based approach.

An evaluation and discussion of the results will be subject to section five. Finally, conclusions are drawn in the last section.

2 State of the Art and Problematic

The recognition, the extraction or/and the classification of NEs from a corpus or a parallel corpora could be done by linguistic approach, statistical one or hybrid approach [19]. [11] proposed an approach for named entity classification using Wikipedia article infoboxes based on attributes builder and a data set of named entities. [4] proposed an approach of recognition and translation based on a representation model of Arabic NEs and a set of transducers resolving morphological and syntactical problems. The main shortcoming of supervised learning described by [13] is the requirement of a large annotated corpus. Interesting results are obtained by using semi-supervised or unsupervised learning methods. [14] proposed a semi-supervised learning for multilingual NE recognition from Wikipedia. Two alignment discriminating models are proposed by [1]. The first model formalizes the alignment operation as a multi-class classification task and processes it with a maximum entropy classifier; the second model is based on Conditional Random Fields (CRF). [15, 19] propose a combination of both approaches in which linguistic analysis is carried out before the application of statistical measures.

Matching single words and nominal phrase from parallel texts is relatively a well controlled task for languages using Latin script but it is complex when the source and target languages do not share the same written script. [16] had proposed an alignment experiment of a Spanish-Arabic Parallel Corpus at the sentence level. The experiment adopts a hybrid methodology, since it applies statistical models, with lexical information (Named Entities as anchor points). [3] proposed a word alignment for translating medical terminologies in a parallel text corpora. [18] used French-Arabic parallel corpora for the study of the impact of proper names transliteration on the quality of the words alignment. [12] studied the problem of matching co-referent named entities by using Soft-TFIDF measure.

[17] has extracted named entities from French-Arabic plants parallel corpora, the objective is to continue this work by matching the extracted named entities in French with the ones in Arabic. Hence, the problematic of this paper is how to use the Latin names of plants for matching the French-Arabic plants common names. Indeed, the scientific name of a plant is a common denominator between the common names because the scientific name in Latin and is independent of other languages. To do this, we will firstly identify and extract the Latin names of plants; then we will model, implement and test a matching approach from the sets: common names of plants of the French corpus (S1), Latin names of the French corpus (S2), common names of plants of the Arabic corpus (S3) and Latin names of the Arabic corpus (S4).

3 Analysis of Latin Plant Name Characteristics

This section is dedicated to the study of the taxonomy of the Latin names of plants, and is strongly drawn from [5, 8] and some specialized organizations such as Royal Botanic Garden Sydney[1], Plant List[2], Master Gardener Program-University of Wisconsin[3], Gardening Know How[4] and Flora Base[5].

Each plant that has been recognized and described has only one valid name, its botanical name. The nomenclature is controlled by the International Association for Plant Taxonomy, which issues an International Code of Botanical Nomenclature that is strictly adhered to throughout the world. The International Code of Nomenclature for Cultivated Plants, which governs the rules for naming cultivars, is issued by the International Union of Botanical Sciences. Both these Codes are revised periodically. The botanical name is the internationally recognized name for a particular plant. Its stem is usually Latin, Greek, or a proper name or descriptive term, and has a Latinized ending. The botanical name consists of two names: the first identifies the genus, and the second (specific epithet) identifies a particular member of the genus. Together the genus and specific epithet constitute the name of the species. The species is the basic unit in a classification system whose members are structurally similar, have common ancestors, and maintain their characteristic features in nature through innumerable generations.

- Cultivar - a man-made species created by hybridization, mutation, or selection for desirable characteristics light color, fragrance, growth habit, and so on.
- Hybrid - an offspring resulting from the breeding of two genetically distinct species. An example of breeding two distinct genera is the cross between Hedera helix and Fatsia japonica. It is written as xFatshederalizei where the "x" is a multiplication sign. An example of breeding between two distinct species is the cross between Plantanus orientalis and Plantanus occidentalis, thus forming Plantanusxacerifolia, the London plane tree. Often, for simplicity's sake, the "x" is removed from the name.
- Variety - a subspecies that is a naturally occurring variation within a species. It is denoted as a word or words, typically in Latin, following the abbreviation "var." A common example is Pittosporum tobira var. variegata. As for "var.", it is often possible to note a common error, as it is left out of the name. A variety is a subdivision of a species, and exhibits various inheritable morphological characteristics (form and structure) that are perpetuated through both sexual and asexual propagation. A variety is designated by a trinomial (three names). The varietal term is written in lower case and underlined or italicized. It is sometimes written with the

---

[1] https://www.rbgsyd.nsw.gov.au.

[2] http://www.theplantlist.org.

[3] http://wimastergardener.org.

[4] http://www.gardeningknowhow.com.

[5] https://florabase.dpaw.wa.gov.au.

abbreviation var. placed between the specific epithet and the variety term (for example, Juniperuschinensis var. surge).

The Code also specifies which form the name must take. The following Table 1 summarizes them:

Table 1. Latin plant taxonomy form

Rank	Ending	Examples
Division (Phylum)	-phyta	Pinophyta, Magnoliophyta
Class	-opsida	Pinopsida, Liliopsida, Magnoliopsida
Order	-ales	Pinales, Liliales, Magnoliales
Family	-aceae	Pinaceae, Liliaceae, Magnoliaceae
Tribe	-eae	Pineae, Lilieae, Magnolieae
Genus	A noun	Pinus, Lilium, Magnolia
Species	Depends	Pinusflexilis, Liliumgrandiflorum, Magnolia grandiflora
Variety	Depends	Pinusflexilis var. humilus
Form	Depends	

3.1   The Family Names

The names of families are plural adjectives used as nouns and are formed by adding the suffix -aceae to the stem, which is the name of an included genus. Thus, the buttercup genus Ranunculus gives us the name Ranunculaceae for the buttercup family and the water-lily genus Nymphaea gives us the name for the water lilies. There are eight families for which specific exceptions are provided and which can be referred to either by their longstanding, conserved names or, as is increasingly the case in recent floras and other published works on plants, by their names which are in agreement with the Code. These families and their equivalent names are:

- Compositae or Asteraceae (on the genus Aster)
- Cruciferae or Brassicaceae (on the genus Brassica)
- Gramineae or Poaceae (on the genus Poa)
- Guttiferae or Clusiaceae (on the genus Clusia)
- Labiatae or Lamiaceae (on the genus Lamium)
- Leguminosae or Fabaceae (on the genus Faba)
- Palmae or Arecaceae (on the genus Areca)
- Umbelliferae or Apiaceae (on the genus Apium)

3.2   Generic Names

The name of a genus is a noun, or word treated as such, and begins with a capital letter. It is singular, may be taken from any source whatever, and may even be composed in an arbitrary manner. The genus (plural genera) may be defined as a more or less closely related and definable group of plants comprising one or more species. The unifying characteristic of a genus is a similarity of flowers. A group of closely related genera is called a family. The botanical name of the family is usually recognizable by its -aceae ending. The stem of the name is the name of one of the genera within the family. For example, Cornaceae is the family name in which Comas is a genus. The etymology of generic names is, therefore, not always complete and, even though the derivation of some may be discovered, they lack meaning. By way of examples:

- Portulaca, from the Latin porto (I carry) and lac (milk) translates as 'milkcarrier'.
- Pittosporum, from the Greek, pitta (tar) and sporoj (a seed) translates as 'tarseed'.
- Hebe was the goddess of youth and, amongst other things, the daughter of Jupiter.
- Petunia is taken from the Brazilian name for tobacco.
- Tecomais taken from a Mexican name.
- Linnaea is one of the names which commemorate Linnaeus.
- Sibara is an anagram of Arabis.

Aa is the name given by Reichenbach to an orchid genus which he separated from Altensteinia. It has no meaning and, as others have observed, must always appear first in an alphabetic listing.

3.3   Species Names

The name of a species is a binary combination of the generic name followed by a specific epithet. If the epithet is composed of two words they must be joined by a hyphen or united into one word. The epithet can be taken from any source whatever and may be constructed in an arbitrary manner. It would be reasonable to expect that the epithet should have a descriptive purpose, and there are many which do, but large numbers of them, either refer to the native area in which the plant grows or commemorate a person (often the discoverer, the introducer into cultivation or a noble personage).

The epithet may be adjectival (or descriptive), qualified in various ways with prefixes and suffixes, or a noun. Specific epithets which are nouns are grammatically independent of the generic name. Campanula trachelium is literally 'little bell' (feminine) 'neck' (neuter). When they are derived from the names of people, they can either be retained as nouns in the genitive case (clusii is the genitive singular of Clusius, the Latinized version ofl'Écluse, and gives an epithet with the meaning 'of l'Écluse') or be treated as adjectives and then agreeing in gender with the generic noun (SorbusleyanaWilmott is a tree taking, like many others, the feminine gender despite the masculine ending, and so the epithet which commemorates Augustin Ley also takes the feminine ending). The epithets are formed as follows [5]:

- to names ending with a vowel (except -a) or -er is added
  - i when masculine singular
  - ae when feminine singular
  - orum when masculine plural
  - arum when feminine plural

- to names ending with -a is added
  - e when singular
  - rum when plural
- to names ending with a consonant (except -er) is added
  - ii when masculine singular
  - iae when feminine singular
  - iorum when masculine plural
  - iarum when feminine plural or, when used adjectivally
- to names ending with a vowel (except -a) is added
  - anus when masculine
  - ana when feminine
  - anum when neuter
- to names ending with -a is added
  - nus when masculine
  - na when feminine
  - num when neuter
- to names ending with a consonant is added
  - ianus when masculine
  - iana when feminine
  - ianum when neuter

When an epithet is derived from the name of a place, usually to indicate the plant's native area but also, sometimes, to indicate the area or place from which the plant was first known or in which it was produced horticulturally, it is preferable to be adjectival and takes one of the following endings:

- -ensis (m) -ensis (f) -ense (n)
- -(a)nus (m) -(a)na (f) -(a)num (n)
- -inus (m) -ina (f) -inum (n)
- -icus (m) -ica (f) -icum (n)

After characterizing the Latin plant name, we present in the next section our proposed method, for the recognition of Latin plant name from French-Arabic corpora and its use for matching common names of plants.

4 Proposed Method and Implementation Using NooJ Platform

Wikipedia is used by many computer linguists applications as varied as construction of ontologies and taxonomies, semantic disambiguation, lexicology and translation [6, 14]. [17] recognized and extracted common plants named entities from French-Arabic parallel corpora built using Wikipedia. Our approach will use these results.

The approach illustrated in Fig. 1 is based on two major steps:

- The first step is to identify the Latin names of plants from the French Arabic corpora. For this purpose, a set of dictionaries and syntactic grammars have been developed and implemented on the NooJ linguistic platform [20]. The linguistic platform NooJ allows the developer to construct, test, and maintain large coverage lexical resources [2, 7, 9], as well as apply morpho-syntactic tools for Arabic processing [4, 10, 20].

- The second step consists of matching the identified plants named entities. This step will use four sets: all the Latin names of the French corpus, all the Latin names of the Arabic corpus which are the results of the previous stage; all named entities of French plants named entities and all Arabic plants named entities from [17]. Using the fact that the Latin name is regardless of the language, the same, we propose a matching based on the Latin plant name.

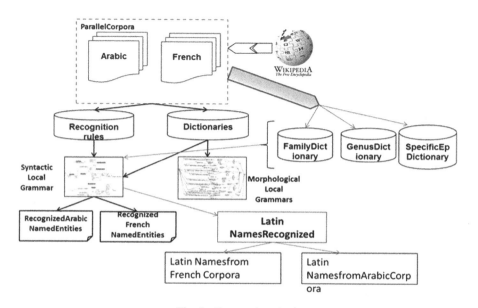

Fig. 1. Proposed method

4.1   Dictionaries

The following dictionaries are created according to the Latin plants names aforementioned morphological analysis.

- A dictionary of Family plants names Fig. 2: it contains 642 entries.
- A dictionary of Genus Fig. 3: it contains 9857 entries.
- A dictionary of Epithets Fig. 4: it contains 1975 entries.
- A dictionary of Variety: it contains 1326 entries.

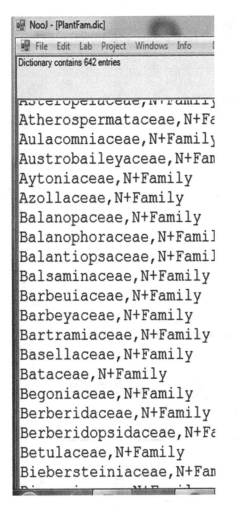

Fig. 2. An extract of family plant dictionary   **Fig. 3.** An extract of genus plant dictionary

4.2   Transducers Implemented on the NooJ Platform

From the Latin plant's name structure, a set of rules, terms and context are identified for modeling the patterns. A set of patterns are modeled and translated into a grammar based on a list of syntactical and morphological rules.

Essentially, the patterns take in consideration Families, Genera and Epithets, because they are the most used plant's scientific name taxonomy. Some cases of varieties are also identified. The grammars perform recognition and extraction of Latin plants names from the input corpus.

The main transducer of Fig. 5 allows recognition of Latin plants named entities belonging to the French/Arabic corpus. Each path of each sub-graph represents a set of rules extracted in the studied corpus.

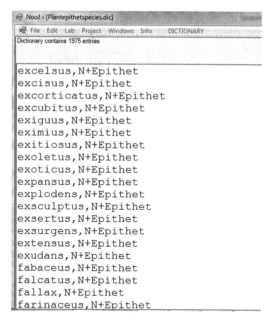

Fig. 4. An extract of epithet plant dictionary

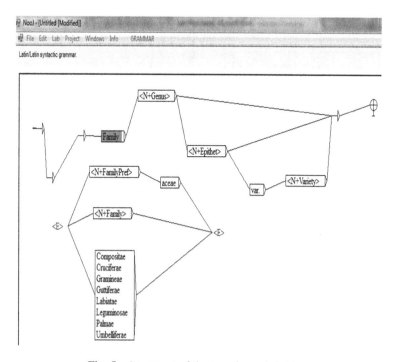

Fig. 5. An extract of the transducer definition

4.3   Matching Approach

The matching approach is based on the recovery from the corpora of the position of
each:

- French plant NE (P_i)
- Latin NE of French Corpora (P_j)
- Arabic plant NE (P_k)
- Latin NE of Arabic Corpora (P_l)

With, $P_j > P_i$ and $P_l > P_k$, because the Latin name both in French or Arabic corpus,
comes after the common name for indexing it. Then, we compute the distances from
the Latin Name according to the rule: when the Named Entities are closer to the same
Latin Name, they are matched.

Figure 6 shows the French plant name Chèvrefeuille des bois followed by the Latin
name LoniceraPericlymenum and the Arabic plant name العسلة الأوروبية al `asalah
'al' uwruwbbiyyah is linked to the same plant name. The proximity of common
names is calculated using the Latin name as a reference point. We avoid to use the
brackets, because some texts do not use them or use comma or add other description
text (Fig. 6 Arabic description).

العسلة الأوروبية (باللاتينية: Lonicera periclymenum) نبات زينة متسلق ينتمي النبات إلى جنس العسلة من
الفصيلة الخمانية. موطنه الأصلي أوروبا من جنوبها إلى شمالها حتى النرويج وبريطانيا. ينتشر حالياً في أمريكا
الشمالية أيضًا.

Le **Chèvrefeuille des bois** *(Lonicera periclymenum)* est une liane
arbustive de la famille des Caprifoliacées aux fleurs très odorantes.

Fig. 6. An extract of a plant's name with their latin name

5   Evaluation Method and Results

[21] presents a systematic analysis of different performance measures. We use the
following measures for evaluating and comparing the results: Precision, Recall and the
F-measure. Precision indicates how many of the extracted entities are correct. Recall
indicates how many entities among those to be found, are effectively extracted.

	Arabic corpus	French corpus
Number of Text	77	77
Number of words	189745	397430
Precision	0,99	0,99
Recall	0,89	0,89
F-measure	0,93	0,93

The lack of noise is due to the characteristic of the well formalized Latin plants names, the extracted Latin name are correct. The silence problem is mainly caused by the lack of some Latin plants names in the dictionaries.

For the matching system we have used the Arabic-French extracted plants named entities. We report the performance of our mapping approach.

Precision	Recall	F-Measure
0,93	0,82	0,87

The precision of the matching is good because the use of the Latin name as "common denominator" for its common names neighbors give a high probability of correct matching. But some cases present a bad matching and this occurs when the Latin name is used in the text as a subject and not for indexing a common name.

The problem for the recall comes essentially from two causes. The first one is that the common named entity is described without a Latin Name, in Fig. 7, the plant name الطيون البانونجي elTayuwn 'albaanuwjy' is without its Latin name, and in that case there is not a matching.

L'**herbe aux mouches** ou **œil de cheval** (*Inula conyza*), encore appelée **inule squarreuse**, est une plante herbacée vivace de la famille des

الطيون الباتونجي[1] نوع نباتي ينتمي إلى جنس الطيون من الفصيلة النجمية.

موطنه أوروبا وآسيا.

Fig. 7. Case of lack of Latin name

The second cause is when a series of synonyms designated by the same Latin name like in Fig. 8, the matching occurs between the last one in the series, in our case sesame with جلجلان jiljlaan, the common names سمسم simsim and زلنجان zilinjaanare not matched.

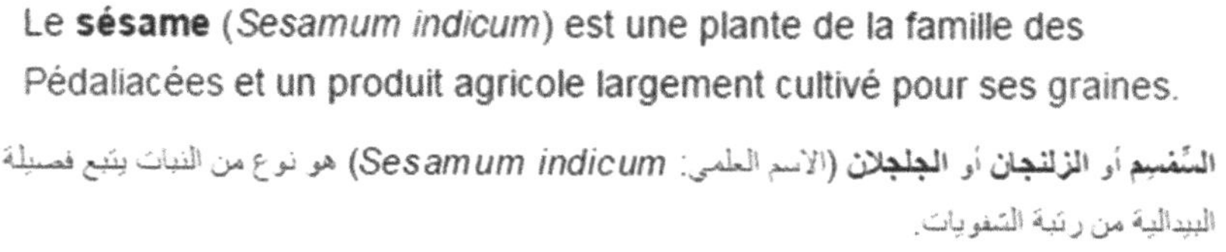

Le **sésame** (*Sesamum indicum*) est une plante de la famille des Pédaliacées et un produit agricole largement cultivé pour ses graines.

السِّمسِم أو **الزلنجان** أو **الجلجلان** (الاسم العلمي: *Sesamum indicum*) هو نوع من النبات يتبع فصيلة البيدالية من رتبة السفويات.

Fig. 8. Case of synonym series

6 Conclusion and Perspectives

In this paper, we first presented a process of recognition and extraction of Latin plants names from French-Arabic parallel corpora. Modeling the recognition phase using patterns with the NooJ platform has yielded very satisfactory results, boosted by the well formalization of the Latin names of plants.

Then, the obtained results of Latin plant named entities were used for the matching of common names of plants. The approach of matching common names of plant in French with their corresponding Arabic using Latin names of plants as a facilitator of the combination has yielded encouraging results. It is true that the majority of texts on plants may not contain the Latin plants names, but this work can be improved and combined with learning algorithms (Expectation-Maximization, k-nearest neighbor, and so on) to generate a bilingual dictionary on plants.

References

1. Allauzen, A., Wisniewski, G.: Modèles discriminants pour l'alignement mot à mot. TAL, vol. 50, no. 3/2009, pp. 173–203 (2009)
2. di Buono, M.P.: Semi-automatic indexing and parsing information on the web with NooJ. In: Okrut, T., Hetsevich, Y., Silberztein, M., Stanislavenka, H. (eds.) NooJ 2015. CCIS, vol. 607, pp. 151–161. Springer, Heidelberg (2016). doi:10.1007/978-3-319-42471-2_13
3. Deléger, L., Merkel, M., Zweigenbaum, P.: Translating medical terminologies through word alignment in parallel text corpora. J. Biomed. Inf. **42**, 692–701 (2009)
4. Fehri, H.: Reconnaissance automatique des entités nommées arabes et leur traduction vers le français. Ph.D. Thesis, Sfax University, Tunisia (2011)
5. Gledhill, D.: The Names of Plants. Cambridge University Press, Cambridge (2008)
6. Goldman, J.-P., Scherrer, Y.: Création automatique de dictionnaires bilingues d'entités nommées grace à Wikipédia. Nouveaux cahiers de linguistique française **30**, 213–227 (2012)
7. Chadjipapa, E., Papadopoulou, E., Gavrillidou, Z.: Adjectives in Greek NooJ module. In: Monti, J., Silberztein, M., Monteleone, M., di Buono, M.P. (eds.) Formalising Natural Languages with NooJ 2014, pp. 49–54. Cambridge Scholars Publishing, Newcastle (2015)
8. Hoffmann, D.: Medical Herbalism: The Science and Practice of Herbal Medicine. Healing Arts Press, Rochester (2003)
9. Maisto, A., Guarasci, R.: Morpheme-based recognition and translation of medical terms. In: Okrut, T., Hetsevich, Y., Silberztein, M., Stanislavenka, H. (eds.) NooJ 2015. CCIS, vol. 607, pp. 172–181. Springer, Cham (2016). doi:10.1007/978-3-319-42471-2_15
10. Mesfar, S., Bessaies, E.: Automated document classification and event extraction in standard Arabic. In: Monti, J., Silberztein, M., Monteleone, M., di Buono, M.P. (eds.): Formalising Natural Languages with NooJ 2014, pp. 236–247. Cambridge Scholars Publishing, Newcastle (2015)
11. Mohamed, M., Oussalah, M.: Identifying and extracting named entities from wikipedia database using entity infoboxes. Int. J. Adv. Comput. Sci. Appl. **5**(7), 164–169 (2014)
12. Moreau, E., Yvon, F., Cappé, O.: Robust similarity measures for named entities matching. In: Coling 2008 Organizing Committee (eds.) Proceedings of the 22nd International Conference on Computational Linguistics, pp. 329–336. Coling, Manchester (2008)

13. Nadeau, N., Sekine, S.: A survey of named entity recognition and classification. In: Satoshi, S., Ranchhod, E. (eds) Lingvisticæ Investigationes, vol. 30, no. 1, 2007, pp. 3–26. John Benjamins Publishing Company, Amsterdam (2009)
14. Nothan, J., Ringland, N., Radford, W., Murphy, T., Curran, J.R.: Learning multilingual named entity recognition from Wikipedia. Artif. Intell. **194**, 151–175 (2013)
15. Ozdowska S., Claveau V.: Inférence de règles de propagation syntaxique pour l'alignement de mots. TAL, ATALA, vol. 47, no. 1, pp. 167–186 (2006)
16. Samy, D., Moreno Sandoval, A., Guirao, J.M.: An alignment experiment of a Spanish-Arabic parallel corpus. In: Proceedings of the International Conference on Arabic Language Resources and Tools (NEMLAR 2004), pp. 85–89 (2004)
17. Seideh, M.A.F., Fehri, H., Haddar, K.: Named entity recognition from Arabic-French herbalism parallel corpora. In: Okrut, T., Hetsevich, Y., Silberztein, M., Stanislavenka, H. (eds.) NooJ 2015. CCIS, vol. 607, pp. 191–201. Springer, Cham (2016). doi:10.1007/978-3-319-42471-2_17
18. Semmar, N., Houda S.: Etude de l'impact de la translittération de noms propres sur la qualité de l'alignement de mots à partir de corpus parallèles français-arabe. In: TALN (eds.) Actes de la 21e conférence sur le Traitement Automatique des Langues Naturelles (TALN 2014), pp. 268–279. Association pour le Traitement Automatique des Langues, Marseille (2014)
19. Semmar, N., Servan, C., De Chalendar, G., Le Ny, B.: A hybrid word alignment approach to improve translation lexicons with compound words and idiomatic expressions. In: ASLIB (eds.) Proceedings of the 32nd Translating and the Computer Conference. ASLIB, London (2010)
20. Silberztein, M.: Formalizing Natural Languages: The NooJ Approach. Wiley, London (2016)
21. Sokolova, M., Lapalme, G.: A systematic analysis of performance measures for classification tasks. Inf. Process. Manage. **45**, 427–437 (2009)

Generating Alerts from Automatically-Extracted Tweets in Standard Arabic

Hiba Chenny[1(✉)] and Slim Mesfar[2]

[1] ESC Tunisia / IHEC Carthage, Manouba, Tunisia
hiba.chenny@gmail.com
[2] RIADI, University of Manouba, Manouba, Tunisia
mesfarslim@yahoo.fr

Abstract. Due to their wide popularity and easy access to the published contents, social media such as Facebook and Twitter have attracted the interest of media to disseminate their information (information, news, events ...). Nowadays, we are witnessing a much-accelerated rhythm of events shared on social media. These events are covering several topics like politics (e.g. presidential elections), epidemics (Zika virus), terrorism (DAECH attacks) or economy (Stock market). Given the importance and relevance of the shared information, several methods and tools are developed to detect and display information from social networks especially Twitter such as MABED1 [1], Twitter Monitor [2] and [3]KeySEE [3].

Keywords: Tweet · NLP · Generating alerts · Twitter · Monitoring system · NooJ · Arabic language

1 Introduction

This work will focus on the automatic extraction of the posted data on Twitter by media in Arabic language (like France 24, El Jazeera, BBC, and so on), and then it will extract events using NooJ's local grammars. Our main goal is to create a monitoring system allowing us a real time control of the evolution of topics that interest people in the world. Our research will be divided into three parts:

- First, the compilation of tweets using a Twitter's API to build a NooJ corpus;
- Then, the extraction of events from the data collected using a linguistic approach based on the NooJ's linguistic engine [11]. These events will be classified into more than 20 domains;
- Finally, the alerts generation of the developed system when abnormal scores are noticed in a given domain.

The Web evolution has brought a new generation of sites facilitating interactions between cybernauts, known as social media. By the year 2006, a new social media has been created by Jack Dorsey named Twitter. This service, has readily gained worldwide popularity. Today, Twitter is one of the most used social media in the world. It allows users to publish short messages limited to 140 characters called tweets. Every second, hundreds of millions of tweets are posted from official accounts as well as personal

© Springer International Publishing AG 2016
L. Barone et al. (Eds.): NooJ 2016, CCIS 667, pp. 145–154, 2016.
DOI: 10.1007/978-3-319-55002-2_13

accounts. These tweets or data carry away many topics and information in real time which can be related to public events, politicians, celebrities, and so on. Considered as the phenomenon of the decade, social networks, especially Twitter, has attracted the interests of media in order to spread their information.

Online press publishes constantly millions of information and a huge volume of tweets on unrelated subjects overwhelms events, on different aspects of life in real time, which make detecting a particular theme of events a difficult task, as the data looked for. An event is commonly defined as something that occurs in a certain place during a particular interval of time.

This paper develops a system that generates alerts about terrorism acts in Twitter's content. To do that, a NooJ's linguistic engine will be used.

This paper has been divided as follow: The first part deals with the theoretical dimensions of the work. The second part shows some previous studies that are related to our research in order to specify the proposed approach. The third part starts with a description of the adopted approach, the used resources as well as the developed grammar. The fourth part describes the evaluation and synthesis of our monitoring system and some of challenges that have been faced. Finally, we discuss the results and perspectives of our research.

2 Related Work

Event detection has become a very active area, found in several scientific works, among which the work of Adrien Guille and Cecile Favre entitled "Event detection, tracking and visualization in Twitter: A mention-anomaly-based approach" [1].

"Event detection, tracking and visualization in Twitter" is a novel-statistical method that relies solely on tweets and leverages the creation frequency of dynamic links that users insert in tweets to detect significant events and estimate the magnitude of their impact over the crowd.

In 2010, Michael Mathioudakis and Nick Koudas [2] developed a system that performs trend detection over the Twitter stream. In addition, in 2013, Pei Lee, Laks V.S. Lakshmanan and Evangelos Milios developed the KeySEE system.

The KeySEE system transforms the social stream into an evolving network post and uses density-based clustering to identify the events.

3 Implementation

3.1 Corpora

This section aims at building a tweet's corpora. For this purpose, these three steps need to be followed: Collecting tweets, generating XML files and developing a filter (Fig. 1).

To collect tweets, we relied on a twitter API (rest API) in order to extract tweets (Fig. 2) in standard Arabic, posted by already-selected official media's account, such as BBC Arabic, Al Jazeera Arabic, Al-Arabiya and CNN Arabic.

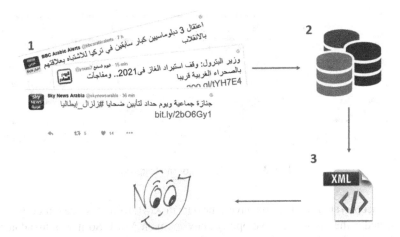

Fig. 1. Corpora construction process

Fig. 2. Main interface of "Collect tweets"

Our choice was based on the number of the followers, which indicates its importance among the cybernauts.

As a first step, these data were stored in a database and organized successively by:

- Time: indicates the date of posting or reposting the tweet under the format dd/mm/yyyy
- Tweet: indicates the data posted,
- Tweeted: indicates the original source of the post
- Screen name: indicates the account that has posted or re-tweeted the tweet.
- Date: indicates the tweet's posting date, under the format yyyy-mm-dd. (Fig. 3).

	id	Time	Tweet	Tweeted	Screenname	date
☐ 🖉 Modifier ᠄ Copier ⊜ Effacer	276	21/02/2016	"ف المهاجرين في ملجأ النار اشتعال بعد "مارة اللهم...	بي بي سي BBC Arabic	BBCArabic	2016-02-21
☐ 🖉 Modifier ᠄ Copier ⊜ Effacer	277	21/02/2016	الرجل جوبي انفجرات سلسلة في "قتلاً 30" | بالفيديو...	بي بي سي BBC Arabic	BBCArabic	2016-02-21
☐ 🖉 Modifier ᠄ Copier ⊜ Effacer	278	21/02/2016	#تحطل يؤدي مبارة لحب مبارة ألف 59 سحب فوتو | اقتصاد...	بي بي سي BBC Arabic	BBCArabic	2016-02-21
☐ 🖉 Modifier ᠄ Copier ⊜ Effacer	279	21/02/2016	#مستخدمين سيارتين فجر في "قتلاً 46" | بالفيديو...	بي بي سي BBC Arabic	BBCArabic	2016-02-21
☐ 🖉 Modifier ᠄ Copier ⊜ Effacer	280	21/02/2016	https://t.co/u... ديها الطوارئ حالة يُمدّ #تونس رئيس	بي بي سي BBC Arabic	BBCArabic	2016-02-21
☐ 🖉 Modifier ᠄ Copier ⊜ Effacer	281	21/02/2016	#القتلة نهمة عاجلة لمحاكمة مصري تحرّض إحالة | مصر...	بي بي سي BBC Arabic	BBCArabic	2016-02-21

Fig. 3. Organization of the Database

In the second step, we had to convert the data in XML format, since our corpora will be built under the linguistic development environment NooJ. So it was mandatory to discard the database and convert it into a textual format (XML, txt, csv…).

For this purpose, automatically generation of an XML file, simultaneously with the extraction of tweets were added (Fig. 4).

```
▼<source name="cnnarabic">
  ▼<tweet Date="23/07/2016" URL="https://t.co/w4q858tBvT">
    خطوط طيران نيوزيلندا تضيف "نكهة هوليوودية" لتحذيرات السلامة سياحة وسفر شركات طيران
  </tweet>
</source>
```

Fig. 4. Extraction of a tweet under the XML format

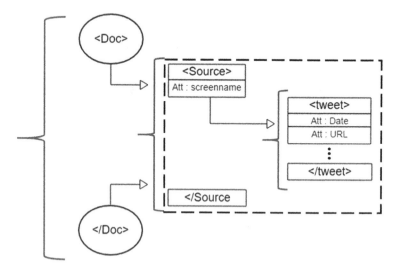

Fig. 5. Tree representation of the XML file

Our XML file can be represented by a tree (Fig. 5), in which we can distinguish two different types of nodes: Main node ("Doc"), and two sub nodes: ("source") having an attribute "name" referring to the screen name and ("tweet") having "Date" as a first attribute referring to the date of the post and a second attribute "URL" referring to its link. The sub node Tweet contains the textual contents.

This solution (conversion into xml file), does not include the tweets that are already stored in our database, hence, the need for developing a filter organized by date and/or the official twitters media's account, as shown in the figure below (Fig. 6):

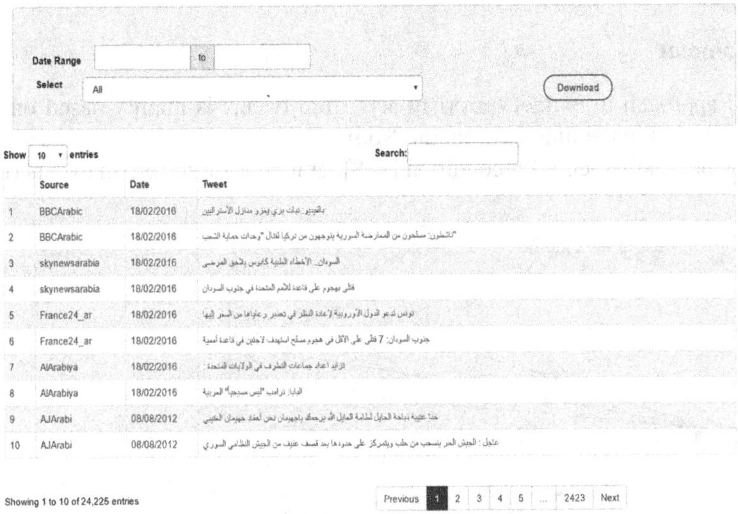

Fig. 6. Filter

The result of these iterations gave an XML file corpus of standard Arabic tweets published by the official media accounts in Twitter. As shown in the figure below (Fig. 7), the corpus contains 171 files, 273 428 222 797 tokens and various annotations. These numbers are incremented with each daily extraction of tweets.

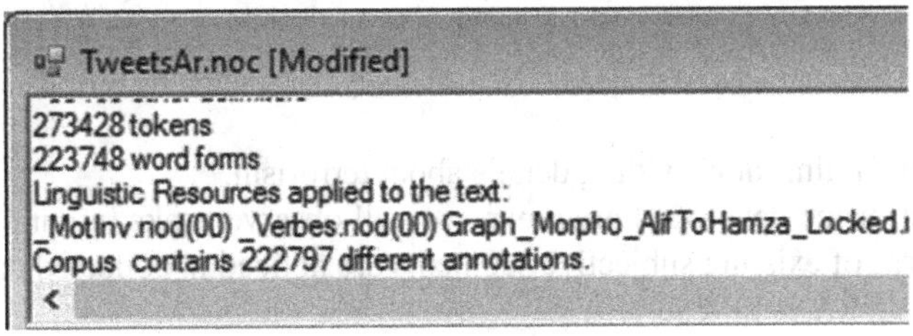

Fig. 7. Results

3.2   Resources

The linguistic engine NooJ will be used to deal with the all morpho-syntactic searches such as lemmatization, annotation and submission of various linguistic queries in order to provide access to advanced features (syntactic and morphological queries …) without being expert in the field. NooJ is a linguistic engine based on large coverage dictionaries and grammars. It parses text corpora made up of hundreds of text files in real time [11]. In addition, the Electronic Dictionary for Arabic "El-DicAr" [10] will be used as the basis for our system.

3.3   Grammar

The used approach to extract terrorism acts from tweets is mainly based on a local grammar created in the linguistic engine NooJ.

Below our Named-entity recognition [7, 8], grammar for terrorism events (Fig. 8):

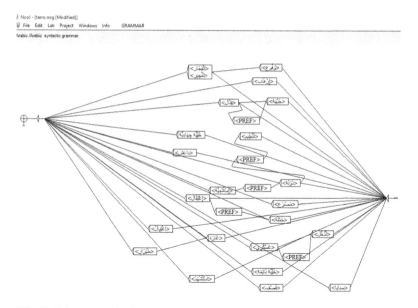

Fig. 8.   Diagram: – NooJ grammar: Named-entity recognition for terrorism event

This grammar aims at providing details about terrorism.

As a result of the developed grammar, we will observe peaks in our system when abnormal scores of existing subjective segments are noticed in collected data.

3.4   Generating Alerts

As mentioned in the first section, the result of our system is a graphical curve with an x and y axis (Fig. 9):

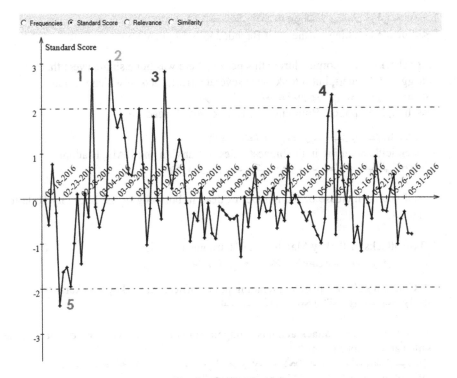

Fig. 9. Generating alerts

The X-axis contains the event's dates under the form M-D-Y. This is justified by the fact that the events must be ordered in chronological order. The second axis described the standard scores that have a determined threshold « 2 » in the both sides positive and negative; once the curve exceeded this threshold, it generates an alert. In this figure above we have 5 alerts. Exploring our corpora, we find that these alerts refer to the events below:

1. <<مقتل 14 شخصا على الأقل بتفجير انتحاري شمال شرقي بغداد>>
   <<Suicide bombing northeast of Baghdad kills 14 people >>

2. By checking our corpus during this period, there was not a single event that triggered the alert, but it took place several terrorist acts of which we quote :
   <<العراق الشرطة العراقية:29ـقتيلافي تفجير جنوب بغداد>>
   << Iraq, Iraqi police: Bombing Baghdad leaves 29 dead >>

   <<تفجير انتحاري في كبرى مدن بابل و داعش يتبنى>>
   <<Suicide bombing in the biggest cities of Babylon and DAECH adopt >>
   <<رويترز : ارتفاع قتلى تفجير مدينة الحلة العراقية إلى 60>>

   <<Reuters: Increase of the number of dead in bombing of the Iraqi city Hilla to 60 >>

3. Two attacks in Turkey March 19, 2016 triggered an alert :
   <<تركيا ... مسلح يقتل شخصين بينهما شرطي في أنقرة>>

   <<Gunman kills two people, including police officer, in Turkey's Ankara>>
   <<4 قتلى على الأقل في تفجير انتحاري وسط اسطنبول>>

   << A suicide bomb attack at a busy shopping area in the Turkish city of Istanbul has killed at least four people>>
   << مقتل وإصابة العشرات في حلب مع استمرار قصف الأحياء السكنية :حلب تحترق سوريا |>>
   <<Worldwide red protest, "Aleppo is burning">>

4. Similarly, we can have alerts in the negative side that signify in this case a period of "peace". It means that, the frequency of the words dealing Terrorism is very low compared with other periods, as shown in picture number 5.

In the following section, we discuss the results of our system and show the faced difficulties, particularly in the Arabic language.

4   Results and Evaluation

A corpus of tweets was automatically collected [5] using the program described above. The studied corpus is composed of a set of journalistic tweets published during the period between 18/02/2016 and 02/10/2016: At this time we loaded 27 419 tweets from different Twitter web media.

Traditionally, the scoring report compares the answer file with a carefully annotated file. The system was evaluated in terms of the complementary precision (P) and recall (R) metrics. Briefly, precision evaluates the noise of a system while recall evaluates its coverage. These metrics are often combined using a weighted harmonic called the F-measure (F).

To evaluate our grammar, we analyze our corpus to extract manually the tweets expressing terrorism. Then, we compare the results of our system with the manually-obtained extraction. The application of our local grammar gives the following result:

Precision	Recall
0.9	0.7

According to these results, the identification of tweets talking about terrorism is acceptable. It is noted that the rate of silence in the corpus is low, which is represented by the recall value 0.70. This is due to the fact that journalistic texts of the corpus are heterogeneous and extracted from different resources. For this reason, an infinite vocabulary expressing terrorism was found (each journalist or author expresses the information in its own way). Another major source of uncertainty, is the absence of some recognition [10] rules of a given structure or word.

Despite the problems described above, the developed method seems to be adequate and shows very encouraging extraction rates. However, other rules must be added to improve the result.

5 Conclusion and Perspectives

It is very important to note that the tweets of our corpus are hetero- generous and from various resources. For this reason, we find a sentence structure infinite expressing terrorists' acts for each (journalist and author) expresses information in its own way. Thus, the syntactic structure of the tweets does not follow a normal sentence structure (a subject group, a conjugated verb and one or more supplements). A tweet can be composed of several hashtags, links; an incomplete sentence followed by a link. The non-recognition is largely due to the lack of rules for recognition of a given structure.

Moreover, in the Arabic language, the ambiguity problem has been posed frequently; this example illustrates a case of semantic ambiguity:

>> في حملة مداهمات واسعة بالجهة...ضبط مخزن متفجرات بالوراق ومطاردات لعناصر إرهابية بأكتوبر<<

In this example, the ambiguity has been reflected in the word "أكتوبر" "October" which means a place, not a month. The use of this word as a place is very rare, which does not favor adding an annotation "place" in the dictionary, because it will aggravate the problems of ambiguity.

A further point is to build event extraction generic grammars in order to handle a max of areas [9], make a thorough study on the semantic analysis to eliminate ambiguities [6] and develop new dictionary resources to detect other important areas as the construction of a sports dictionary.

References

1. Guille, A., Favre, C.: Event detection, tracking and visualization in Twitter: a mention-anomaly-based approach. Soc. Netw. Anal. Min. **5**, 18 (2015). Springer, Cham
2. Mathioudakis, M., Koudas, N.: Twitter monitor. trend detection over the twitter stream. In: SIGMOD/PODS 2010 International Conference on Management of Data, pp. 1155–1158. ACM, New York (2010)
3. Milios, P.L.: KeySee. supporting keyword search on evolving events. In: Dhillon, I.S., Koren. Y., Ghani, R., Senator, T.E., Bradley, P., Parekh, R., He, J., Grossman, R.L. (eds.) KDD 2013 - Proceedings of the 19th ACM SIGKDD International Conference on Knowledge Discovery and Data Mining, pp. 1478–1481. ACM, New York (2013)
4. Boujelben, I.B.H.: Transformational analysis of arabic sentences. application to automatically extracted biomedical symptoms. In: Vučković, K., Bekavac, B., Silberztein, M (eds.) Formalising Natural Languages with NooJ. Selected Papers from the NooJ 2011 International Conference, pp. 182–194. Cambridge Scholars Publishing, Newcastle (2011)
5. Boujelben, I.B.H.: Transformational analysis of Arabic sentences. Application to automatically extracted biomedical symptoms. In: Vučković, K., Bekavac, B., Silberztein, M (eds.) Formalising Natural Languages with NooJ. Selected Papers from the NooJ 2011 International Conference, pp. 182–194. Cambridge Scholars Publishing, Newcastle (2011)
6. Ellouze, S., Engin, H.K.: NooJ disambiguation local grammars for Arabic broken plurals. In: Gavriilidou, Z., Chatzipapa. E., Papadopoulou. L., Silberztein, M. (eds.) Proceedings of the NooJ 2010 International Conference, pp. 62–72. University of Thrace Ed., Greece (2010)
7. Fehri, H., Haddar, K., Ben Hamadou, A.: Automatic Recognition and semantic analysis of Arabic named entities. Applications of finite-state language processing. In Judit, K., Silberztein, M., Varadi, T. (eds.) Selected Papers from the NooJ 2008 International Conference. pp. 101–113. Cambridge Scholars Publishing, Newcastle (2010)
8. Chenny, H.: Recognition of communication verbs with NooJ. In: Koeva, S., Mesfar, S., Silberztein, M. (eds.) Formalising Natural Languages with NooJ 2013, pp. 153–168. Cambridge Scholars Publishing, Newcastle (2014)
9. Najar, S.M.: A large terminological dictionary of Arabic compound words. In: Okrut, T., Hetsevich, Y., Silberztein, M., Stanislavenka, H. (eds.) Automatic Processing of Natural-Language Electronic Texts with NooJ. Springer, Cham (2015)
10. Mesfar, S.: Named entity recognition for Arabic using syntactic grammars. In: Kedad, Z., Lammari, N., Métais, E., Meziane, F., Rezgui, Y. (eds.) NLDB 2007. LNCS, vol. 4592, pp. 305–316. Springer, Heidelberg (2007). doi:10.1007/978-3-540-73351-5_27
11. Silberztein, M.: Formalizing Natural Languages: The NooJ Approach. Wiley, London (2016)

Syntactic Analysis

Detection of Verb Frames with NooJ

Krešimir Šojat[⊠], Božo Bekavac, and Kristina Kocijan

Faculty of Humanities and Social Sciences, University of Zagreb,
Zagreb, Croatia
{ksojat,bbekavac,kkocijan}@ffzg.hr

Abstract. This paper deals with semi-automatic extension of CroDeriV with verb valency frames. CroDeriV is a morphological database of Croatian verbs. In its present shape the database comprises 14 500 verbs in infinitive forms. Each verb in CroDeriV is segmented into lexical and derivational morphemes and verbs of the same root are mutually linked. In order to further enrich the CroDeriV with semantic and syntactic information, we have used the NooJ platform to recognize derivationally related verbs, find the verb frames and to speed up the sentence processing.

Keywords: Verb frames · Nooj · Derivation · Morphological grammars · Croatian

1 Introduction

This paper deals with semi-automatic extension of CroDeriV with verb valency frames. CroDeriV is a morphological database of Croatian verbs. In its present shape, the database comprises almost 14 500 verbs in infinitive forms. Each verb in CroDeriV is segmented into lexical and derivational morphemes and verbs of the same root are mutually linked [7]. The database is available for search at http://croderiv.ffzg.hr/ Croderiv.

The lexicon structured in this way enables the recognition of derivationally related families of verbs and, at the same time, the detection of derivational spans of particular base forms.[1] In the second phase of its development, CroDeriV is extended with definitions of verbal meanings, i.e. verbal lexemes are analyzed for their meaning structure and divided into lexical units. Each lexical unit is accompanied with one or more sentences illustrating its contextual usage. These sentences also function as a basis for the construction of valency frames, i.e. frames reflecting verbal argument structure.

In this paper we focus on the detection of verbal arguments in sentences using NooJ. We also discuss the structure of existing morphological grammars for Croatian verbs as well as their future development. We experiment with small derivational families of verbs consisting of 4 and 5 derivative forms around a central member of the family – a base form. Such relatively small derivational families enable a careful design of rules

---

[1] CroDeriV resembles databases like CatVar for English and Uni-morph for Russian (http://courses. washington.edu/unimorph and http://clipdemos.umiacs.umd.edu/catvar).

© Springer International Publishing AG 2016
L. Barone et al. (Eds.): NooJ 2016, CCIS 667, pp. 157–168, 2016.
DOI: 10.1007/978-3-319-55002-2_14

or a redesign of already existing ones. The final goal is to detect major constituents in sentences and automatically classify them according to their syntactic function.

The paper is structured as follows: in Sect. 2 we present the design and structure of CroDeriV in its present form. Section 3 deals with the expansion of CroDeriV with valency frames. In Sects. 4 and 5 we describe the detection of derivationally related verbs and verb frames using NooJ. Section 6 briefly describes the corpus used for this purpose and presents some comparative results for two approaches to verbal derivations. The final part of the paper provides an outline for future work.

2   CroDeriV

CroDeriV is a computational lexicon that provides information on the morphological structure of approximately 14 500 Croatian verbs collected from different sources, mainly digital and paper dictionaries [2, 9] and additionally enriched with lemmas from the Croatian National Corpus v3.0 [10] and the Croatian Web Corpus v2 [4].

The Croatian language has a very rich morphology, both in terms of derivation and inflection. Further in this paper we focus on derivational relatedness of verbs, particularly on the derivation of verbs from other verbs. We also discuss how the lexical entries for derivationally related verbs can be extended with data on their argument structure (valency) and morphosyntactic features of arguments. The general purpose of CroDeriV is to obtain a complete morphological analysis of Croatian vocabulary. At its current stage of development, this resource contains only verbal lemmas, whereas the extension with other parts of speech is planned in future development.

Verbs in Croatian can be derived from other verbs via two derivational processes – prefixation and suffixation. Prefixation is far more productive than suffixation. Prefixes are always derivational, whereas suffixes can be derivational as well as inflectional. In the majority of cases, base forms take one prefix. However, one verbal root as a part of a base form can in some rare cases co-occur even with four prefixes. As far as suffixes are concerned, one root usually has two derivational and one inflectional suffix (-ti). This structure can be extended with an additional suffix that denotes a diminutive or pejorative action. Verbs in Croatian can also be formed by compounding, i.e., they can consist of two roots. Although, compounding is not a very productive process as far as Croatian verbs are concerned.[2]

The morphological structure of all verbal lemmas in CroDeriV was determined in several steps. Due to numerous phonological changes, all prefixes were manually analyzed and segmented, whereas the suffixal part was in the first step segmented automatically. However, the results of automatic rule-based segmentation were not satisfactory. The two main problems in the automated processing of Croatian derivation are homography that results in the overlapping of prefixes and suffixes with roots, and numerous phonological changes at the morpheme boundaries resulting in several allomorphs for each morpheme. All results of automatic segmentation were therefore manually checked and all allomorphs were connected to their unique representative morphemes.

---

[2] There are approximately 120 verbal compounds recorded in CroDeriV.

Lexical entries in CroDeriV consist of verbs decomposed into morphemes and linguistic metadata. The metadata in lexical entries indicate verbal aspect and types of reflexivity. The structure provided for all analyzed verbs consists of 11 morpheme slots and covers all combinations of recorded lexical and grammatical morphemes.[3] There are four types of slots for morphemes: (1) derivational prefixes (four slots), (2) roots (three slots – in the majority of cases only one is filled, the three slots are provided for verbal compounds of two roots and an interfix), (3) derivational and conjugational suffixes (three slots), and (4) infinitive ending (one slot).

This kind of processing enables the recognition of all allomorphs of a particular morpheme and the detection of all affixes co-occurring with particular roots. This procedure also enables the detection of complete derivational families of Croatian verbs. A derivational family consists of verbs with the same lexical morpheme grouped around a base form. Generally, a verb with the simplest morphological structure serves as a base form for verb-to-verb derivation. For example, there are four derivatives recorded in CroDeriV of the base form *jedriti* (to sail). These are *do-jedriti* 'to sail to', *od-jedriti* 'to sail away', *pre-jedriti* 'to sail across' and *za-jedriti* 'to start sailing'. This group of verbs constitutes a derivational family consisting of five members. As indicated, all derived forms in this family are produced through prefixation. In Fig. 1 we present the morphological segmentation of this derivational family as structured in CroDeriV.

Fig. 1. An example of morphological segmentation and derivational relatedness of verbs from CroDeriV.

3   Verb Frames

As mentioned in Introduction, CroDeriV is being extended with definitions of verbs' meanings. In other words, verbal lexemes are manually analyzed for their meaning and consequently divided into lexical units. This enrichment of lexical entries in CroDeriV is motivated by two reasons. The first one is to enable an accurate and extensive description of derivational processes in Croatian, which cannot be done if the semantic

---

[3] A more detailed account is given in [6].

component is not taken into account. If we deal only with 'raw' data, as they are presently recorded in CroDeriV, numerous derivational processes cannot be adequately described. For example, the verb *hodati* can have (at least) two senses in Croatian: 'to walk' and 'to date somebody'. The derivative *prohodati* 'to start walking' is semantically related to both of them, but the derivative *prehodati* 'to cover a certain distance by walking' is semantically related only to the first sense. Without definitions of meaning more comprehensible studies in this area are not possible.

Secondly, the introduction of definitions of meaning and the division of verbal lexemes into corresponding lexical units can also enable the development of a large-scale verb valency description to be used in various NLP tasks. As in previous examples, different senses of the same lexeme *hodati* have different valency or argument structures. The lexeme *hodati* is divided into lexical units that correspond to detected senses of verbal lexemes (e.g. *hodati* 1 – to walk; *hodati* 2 – to date somebody). The division of lexemes into lexical units is based on the analysis of sentences from available corpora, mainly the Croatian National Corpus. Each lexical unit is accompanied with one or more sentences illustrating its contextual usage. These sentences also function as a basis for the construction of verb valency frames. Each frame contains information on argument structure characteristic for a particular lexical unit. In the next step of analysis, the arguments are annotated in contextual examples. On this level of processing their syntactic functions and morphological features are annotated in sentences (e.g. SUB/NOM – subject in nominative case; OBJ/PP ACC – prepositional object in accusative).

We believe that the derivational lexicon of verbs enhanced with the data on their valency can provide a valuable information for various linguistic studies and the development of NLP applications. In Fig. 2 we provide an example of an entry for the lexical unit *gledati* 'to perceive by sight'.

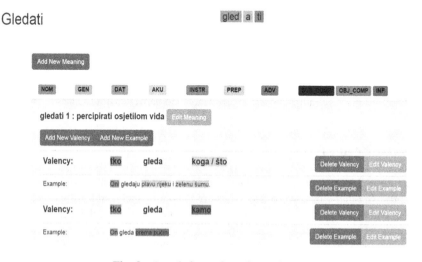

Fig. 2. A verb frame from CroDeriV.

For the visual presentation of sentential elements we use different colors for annotated elements. These elements mainly correspond to constituents. However, the whole procedure is a very time-consuming work if performed completely manually. In order to speed up the building and the development of verb frames we decided to use NooJ. The detection of derivationally related verbs and verb frames with NooJ is described in the following sections.

4 Detection of Derivationally Related Verbs with NooJ

In our research, we decided to use NooJ [5] as an NLP tool since it allows an easy construction of grammars on both morphological and syntactic levels. We needed both in order to recognize derivationally related families of verbs (morphological grammar) before we are able to detect verb valency frames (syntactic grammar). In this chapter, a more detailed description of the morphological grammar will be given and the reasons behind this grammar will be explained. In the next chapter we will do the same for the syntactic grammar while more information on detection of syntactic verbal frames for Croatian using NooJ can be found in [3].

Croatian dictionary of verbs consists of 3 907 verbs (main lemmas). Each verb has a category attribute 'V' and a link to its inflectional paradigm 'FLX = V_PARADIGM'. So far, there are 239 unique inflectional paradigms that combined with the main lemmas produce 342 292 different inflectional forms for verbs (including only simple verb tenses). There are 464 verbs in the dictionary that may appear as reflexive verbs (*ponašati se* 'to behave oneself' or *ponašati se* 'to behave or act'). They are marked with an additional attribute value set '*Prelaz = pov*'. This feature is used in the syntactic grammar for the recognition of compound verb forms (future tenses, past tenses and so on) since an auxiliary and main verb forms may be disconnected with the particle '*se*' denoting the reflexivity of the verb. Recently, we have annotated 91 verbs of perception as the '*prcp*' verbs with additional marking of the perception type (*viz*-vision, *miris* - smell, *sluh* - sound, *okus* - taste and *dodir* - touch). There are also some verbs in our dictionary that have been annotated with their valency frames but this includes only 102 verbs of consummation [8] and 1 739 verbs that were considered the most frequent ones [13].

The existing dictionary is far from being a complete one, and it is clear that it does not hold all of the Croatian verbs. For example, although the verb *bližiti* 'to come near' is in the dictionary, the four verbs that belong to the same family *približiti* 'to bring closer', *zbližiti* 'to make a bond', *približavati*, and *zbližavati* are not. Examples like these hold true for numerous derivational families. Luckily, due to the morphological rules used to build derivationally related verbs, it is quite obvious that an implementation of some types of rules should change this situation and automatically 'populate' the list of (missing) verbs. The real question was to choose the right path (for us) that will enable this project. In fact, NooJ offers two solutions for recognizing derivations. One is via direct input of an attribute +DRV next to the dictionary entry and the other via morphological grammar(s). We will try them both and give our pros and cons for both approaches.

The first approach requires that an attribute-value (+DRV=*derivationName*) combination is inserted in-line for each verb in the dictionary and that all *derivationName* values are defined in a separate NOF file (similarly as it is done for +FLX attribute of an each verb). The average number of derivations is 4–5 per verb, but some verbs may take from two to up to even thirty+derivations. The approximate time for adding a single derivation is 10 min (including the time to determine the correct derivational paradigm and define it in the dictionary)[4]. Considering the present number of verbs in our dictionary, this adds up to 39 070 min or 81 days (8 h/day) of work. Still, we will not be sure if all the possible derivations for each verb are described since some are rarely used and found only in specific genres. On the other side, regardless the (occasional) smaller recall, this approach would provide us with a higher precision.

The second approach gives us all the possible derivations much faster (only 2 days (8 h/day)). In spite of the high recall, it gave us a somewhat lower precision than we have hoped for, but enough data that we can learn from in order to improve the grammar and raise the precision. The grammar's detailed description follows, since it is the path we have chosen for this research.

Our morphological grammar has an L1 priority level, which means that it is applied to the text only after all other morphological resources and only to the words not recognized until that point. The grammar's transducer adds the main verb (the one that the new verb is derived from) in the output as the super-lemma, and it uses its syntactic and inflectional information where applicable. The grammar recognizes, not only the main lemma, but all its gender, number and tense dependent endings, i.e. forms.

The new verbs fall into three main categories:

A. verbs built only with a prefix,
B. verbs built only with a suffix,
C. verbs built with a prefix and a suffix.

Thus, our grammar also has three main branches responsible for the recognition of each type (Fig. 3).

Firstly, we built the branch for recognizing the type A verbs. The branch finds the predefined prefix and then checks if the remaining of the word is equal to any existing[5] verb form. If such a word is found, the transducer marks the new word as a verb that inherits all the features of the verb it is derived from. Thus, the new word is immediately annotated with the gender, number and tense information. For the purposes of this research, we have added some additional information that include the type of prefix ('+PREF') and the main verb whose family the new verb belongs to ('+GL').

Secondly, we built the branch recognizing the type B verbs. At first, we used the same methodology as in the type A verbs. We split the word in three sections where the second section is one of the possible suffixes. Everything before that, with the addition of an ending '-ti', must match an existing verb in the infinitive form, and everything after the suffix should be a tense, gender and number ending. However, this reasoning recognized too many false positives mostly nouns or adjectives that have the same

---

[4] It will take even longer if the new paradigm is needed to be defined in the NOF file.

[5] Where existing means previously built by the inflectional grammar of a dictionary entry.

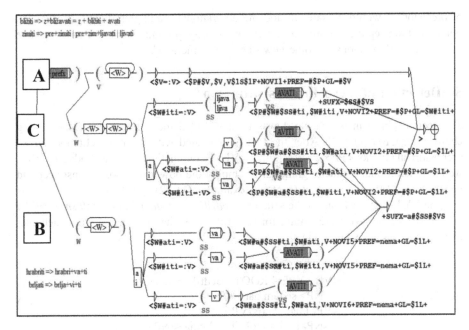

Fig. 3. Morphological grammar for recognizing derivationally related verb families

rooth and the suffix (1st and 2nd sections) but not ending or the 3rd section (for example: *svladavanje* = (svlada+ti) + (va) + (nje) or *čuvarice* = (ču+ti) + (va) + (rice)).

At the end, we built the branch recognizing the type C verbs that uses the same set of predefined prefixes and suffixes and the remaining of the branch is similar to the type B branch. The main difference between them is in the output form of the recognized word.

The grammar proposed here recognizes, in addition to the verbal derivations, some additional words that share similar structures as the ones described for derivations of verbs. We can classify these false positives into four categories:

- nouns: ex. *paravan* [**para**+va+n] the grammar incorrectly recognizes the root as a verb *para+ti* (en. to tear apart) and *para+va+n* as its derivation (using the branch B)
- adjectives: ex. *ekstra* [eks+**tra**] the grammar incorrectly recognizes the root as a verb *tra+ti* (en. to trifle) and *eks+trati* as its derivation (using the branch A)
- verbs: ex. *šarmirati* [š+**armirati**] the grammar incorrectly recognizes the root as a verb *armirati* (en. to reinforce concrete) and *š+armirati* as its derivation (using the branch A)
- foreign words: ex. *delete* [de+**lete**] the grammar incorrectly recognizes the root as a verb *let+(je)ti* (en. to fly) and *de+lete* as its derivation (using the branch A).

If we are to add the missing nouns, adjectives and verbs into our main dictionary, we would be able to solve this problem and subsequently raise the precision of our proposed model. Still, the problem of foreign words would remain, since we do not have these words in the Croatian dictionary. Thus, to completely deal with the problem of precision, it would be safer to add the derivational paradigms directly to the dictionary but to keep the grammar as a very low priority grammar just in the case we have missed some derivations or some new ones have emerged[6].

5   Detection of Verb Frames with NooJ

The second phase in which NooJ was used included the application of an existing chunker for Croatian [12]. After the text is annotated with the noun chunks <NP> , prepositional chunks <PP> , verb chunks <VP> , conjunctions <C> and adverbs <R>, we apply additional syntactic grammar to employ the power of NooJ's transducer and generate an XML-like notation (sentence examples (1), (2), (3), (4)).

The XML-like notation for the sentence: *Jedrili smo sedam dana, imali smo boljih i lošijih jedrenja* – 'We were sailing for seven days, we had better and worse achievements' is given in example (1).

$$
\begin{aligned}
&\text{<SENTENCE>}\\
&\quad\text{<VP TYPE="ROOT">Jedrili smo</VP>}\\
&\quad\text{<NP> sedam dana </NP>,}\\
&\quad\text{<VP>imali smo</VP>}\\
&\quad\text{<NP>boljih i lošijih jedrenja </NP>.}\\
&\text{</SENTENCE>}
\end{aligned}
\tag{1}
$$

The XML-like notation for the sentence: *Četvrti dan jedrilo se sa Velikog Iža na Veliki Rat* 'On the fourth day we sailed from Veliki Iž to Veliki Rat' is given in example (2).

$$
\begin{aligned}
&\text{<SENTENCE>}\\
&\quad\text{<NP> Četvrti dan </NP>}\\
&\quad\text{<VP TYPE="ROOT">jedrilo se</VP>}\\
&\quad\text{<PP> sa Velikog Iža </PP>}\\
&\quad\text{<PP>na Veliki Rat</PP>.}\\
&\text{</SENTENCE>}
\end{aligned}
\tag{2}
$$

The file generated in this manner allows us to produce visual representations of each root verb[7] environment (Fig. 4) as well as the environment of derived verbs (Fig. 3). The chunks are color coded to ensure faster recognition.

The same XML-like notation is provided for the derived verbs such as for the sentence: *Unatoč lošem ulasku u natjecanje i mnogim ostalim problemima, Fantela I*

---

[6] If we observe language as a living thing, it is quite expected that some new derivations will appear in time.

[7] In this paper, we will refer to the selection of main verbs used in this research as the root verb and they are marked as TYPE = "ROOT", while the derivations of these main verbs are marked as TYPE = "NOVIx". All the other types of verb that may appear in the sentence are only marked as an <VP> chunk with no additional attributes.

Marinić sun a kraju uspjeli dojedriti do bronce 'Despite of a bad start and many other problems, Fantela and Marinić won the bronze medal' (example 3).

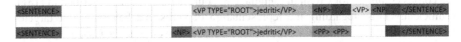

Fig. 4. Example of visual representation of root verb frame

```
<SENTENCE>
    <PP>Unatoč lošem ulasku
      <PP>u natjecanje </PP>
    </PP>
    <C> i </C>
    <NP>mnogim ostalim problemima</NP>,
    <NP>Fantela i Marenić </NP>
    <VP> su </VP>
    <PP> na kraju </PP>
    <VP> uspjeli
      <VP TYPE="NOVI1" PREF="do" GL="jedriti" Verb="dojedriti"> dojedriti
      </VP>
    </VP>
    <PP> do bronce </PP>.
</SENTENCE>
```
(3)

And the sentence: *Još malo i prvi će kišni oblak dojedriti iza obzora i opet ćee padata kiša* – 'Very soon the first cloud will appear on the horizon and it is going to rain again' is presented in example (4).

```
<SENTENCE>
    <R> Još malo </R>
    <C> i </C>
    <NP> prvi </NP>
    <VP> će </VP>
    <NP> kišni oblak </NP>
    <VP TYPE="NOVI1" PREF="do" GL="jedriti" Verb="dojedriti">dojedriti  (4)
    </VP>
    <PP> iza obzora </PP>
    <C> i </C>
    <R> opet </R>
    <VP>će padati kiša</VP>.
</SENTENCE>
```

In Fig. 5 we give a visual representation of verb frames for sentences (3) and (4).

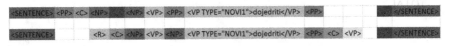

Fig. 5. An example of visual representation of derived verb frame

6 Corpus and the Results of an Experiment

The corpus compiled for Croatian Language Resources for NooJ, as described in [13] is insufficiently large regarding the representation of verbs that belong to our targeted derivational families. An ideal resource for our research would be the Croatian National Corpus (CNC), a representative corpus of the contemporary Croatian standard language. The CNC contains written texts published since 1990. The corpus is automatically lemmatized and MSD tagged using standard heuristic methods as described in [10]. The methodology of coping with unknown words is in more detail explained in [1] together with accuracy statistics which can be relevant for our work.

As CNC comprises over 200 million tokens we decided to extract only the relevant sentences, i.e. the sentences which contain verbs of our potential interest. The extracted subset of CNC in the raw textual format comprises 15 thousand tokens that were imported in NooJ and processed with current version of Croatian lexical module described in [11]. Using such a corpus preparation methodology we deliberately sacrificed some potentially interesting quantitative data about frequencies of certain verbs in the entire CNC. On the other side, we provided focus related corpus optimal for NooJ processing which suits our research needs.

In order to test which approach will score better on the recognition of derived verbs, we have used an even smaller subcorpus of 2 671 tokens. For this experiment, we have used five root verbs and their 4–5 derivations. The Table 1 gives a complete list of the base verbs, derived verbs and the number of their derivations related to gender, number and simple tense that were generated in the inflectional dictionary. This file is automatically produced after joining the dictionary (*.dic) and the morphological derivational description file (*.nof).

Table 1. The number of derivations for the 5 root verbs we have used in the experiment.

Base forms	Derived verbs	Number of derivations
bližiti	*zbližiti, približiti* *zbližavati, približavati*	465
brljati	*zabrljati* *brljaviti, zabrljaviti* *zabrljavati*	243
hrabriti	*ohrabriti, obeshrabriti* *ohrabrivati, obeshrabrivati*	465
jedriti	*dojedriti, odjedriti, prejedriti, zajedriti*	375
kopirati	*fotokopirati, prekopirati, iskopirati, nakopirati*	560
zimiti	*prezimiti, uzimiti, zazimiti* *prezmiljavati* *zimovati*	468
TOTAL		2576

We have tested both approaches and the results (Table 2) clearly show in favor of the first test, i.e. adding derivations directly to the dictionary (via +DRV feature). As expected, the precision is somewhat higher, although still not perfect, in the first test which can be explained with the ambiguity of Croatian language. In our example text, all of the verbs that were incorrectly (or ambiguously) marked as derived verbs, where actually either adjectives (*hr. hrabri* – which may be an adjective 'brave' or a verb 'to encourage') or nouns (*hr. zime* – which may be a noun 'winter' or a verb 'to spend a winter'). However, this ambiguity did not come as a surprise to us, since it is very common in Croatian language, but we hope to resolve it on a syntactic level of analysis, especially with the addition of information on verb valency frames.

Table 2. Comparative results for the two approaches.

Measure	The dictionary approach (+DRV)	The morphological grammar approach
Precision	0,9674	0,9654
Recall	1	0,9398
f-measure	0,9834	0,9524

At this time, we do not offer any statistics on the verb frames since we are still in the process of analyzing and learning from that data.

7 Conclusion

In this paper we have presented a method for the extension and enrichment of a derivational lexicon of Croatian – CroDeriV. In its present form it contains only verbs, whereas other parts of speech will be introduced in future development. CroDeriV is also being extended with definitions of meaning and verb valency information. The project described in this paper has served us as an experiment and a starting point in learning about verb frames with the help of NooJ. As more data will be collected, i.e. more sentences on each verb (root verb and derived verbs), we will be able to define with greater accuracy the exact frame each verb may appear in text. It will take some fine-tuning of the chunker and adding some more identifiers to the chunks in order to make them more informative about the environment the specific families tend to appear in. We believe that this procedure can be valuable for further development of existing language resources like CroDeriV as well as for the development and refinement of existing NooJ resources for the Croatian language.

References

1. Agić, Ž., Tadić, M., Dovedan, Z.: Evaluating full lemmatization of croatian texts. In: Recent Advances in Intelligent Information Systems, pp. 175–184. Academic Publishing House EXIT, Warsaw (2009)
2. Anić, V.: Rječnik hrvatskoga jezika. Novi liber, Zagreb (2004)

3. Bekavac, B.; Šojat, K.: Syntactic patterns of verb definitions in Croatian WordNet. In: Vučković, K., Bekavac, B., Silberztein, M (eds.) International Conference on Formalising Natural Languages with NooJ. Selected Papers from the NooJ 2011, pp. 112–121. Cambridge Scholars Publishing, Newcastle (2011)

4. Ljubešić, N., Erjavec, T.: hrWaC and slWac: compiling Web Corpora for Croatian and slovene. In: Habernal, I., Matoušek, V. (eds.) TSD 2011. LNCS (LNAI), vol. 6836, pp. 395–402. Springer, Heidelberg (2011). doi:10.1007/978-3-642-23538-2_50

5. Silberztein, M.: The NooJ Manual (2003). www.nooj4nlp.net

6. Šojat, K., Srebačić, M., Štefanec, V.: CroDeriV i morfološka raščlamba hrvatskoga glagola. Suvremena lingvistika **39**(75), 75–96 (2013). Zagreb

7. Šojat, K., Srebačić, M., Pavelić, T., Tadić, M.: CroDeriV: a new resource for processing Croatian morphology. In: Calzolari, N., Choukri, K., Declerck, T., Loftsson, H., Maegaard, B., Mariani, J., Moreno, A., Odijk, J., Piperidis, S. (eds.) Proceedingx of the 9th International Conference on Language Resources and Evaluation, pp. 3366–3370. Reykjavik, Iceland (2014)

8. Šojat, K., Vučković, K., Tadić, M. Extracting verb valency frames with NooJ. In: Ben Hamadou, A., Mesfar, S., Silberztein, M. (eds.). International Conference and Workshop on Finite State Language Engineering, NooJ 2009, pp. 231–242. Centre de Publication Universitaire, Sfax, Tunisia (2010)

9. Šonje, J. (ed.): Rječnik hrvatskoga jezika. Leksikografski zavod. Miroslav Krleža & Školska knjiga, Zagreb (2000)

10. Tadić, M.: New version of the Croatian National Corpus. In: After Half a Century of Slavonic Natural Language Processing, pp. 199–209. Masaryk University, Brno (2009)

11. Vučković, K., Tadić, M., Bekavac, B.: Croatian language resources for NooJ. CIT J. Comput. Inf. Technol. **18**, 295–301 (2010). Zagreb

12. Vučković, K.: Model parsera za hrvatski jezik. Ph.D. dissertation. Faculty of Humanities and Social Sciences, Zagreb (2009)

13. Vučković, K., Mikelić Preradović, N., Dovedan, Z.: Verb valency enhanced Croatian Lexicon. In: Judit, K., Silberztein, M., Varadi, T. (eds.): International Conference on Selected Papers from the NooJ 2008, pp. 52–59. Cambridge Scholars Publishing, Newcastle (2010)

Phrasal Verb Disambiguation Grammars: Cutting Out Noise Automatically

Peter A. Machonis[✉]

Florida International University, Miami, USA
machonis@fiu.edu

Abstract. Previous research [1, 2] showed how NooJ could automatically annotate English Phrasal Verbs (PV), both continuous and discontinuous, in large corpora. Due to certain restrictions, however, not all discontinuous PV listed in the PV Dictionary were successfully identified in texts. Further research [3] showed how a simplified PV grammar could identify more PV and improve recall, but it created an excessive amount of noise. Some of it could be automatically removed with disambiguation grammars, yet accuracy was still limited to 70–74%. In this article we show how incorporating additional dictionaries and disambiguation grammars – modifying them with unique NooJ functionalities such as +EXCLUDE and +UNAMB – can allow us to remove even more noise and achieve a better overall accuracy of 88%.

Keywords: NooJ · Natural Language Processing · Multiword expressions · English Phrasal Verbs · Particles · Prepositions · Disambiguation · Historical Linguistics · Dickens · Melville

1 Introduction

Machonis [1, 2] previously showed how NooJ could easily annotate English Phrasal Verbs (PV) in large corpora, with 84% accuracy, including discontinuous PV involving insertions of three or more word forms, such as:

1. That **turned around** the national economy.
2. That **turned** the national economy **around**.
3. That program **brought down** our crime rates.
4. That program **brought** our crime rates **down**.
5. Max **turned on** the computers.
6. Max **turned** the computers **on**.
7. Mayor Ed Koch **has out** a great new book.
8. Mayor Ed Koch **has** a great new book **out**.

Most of our testing has been on 19th century novels, such as Henry James' *Portrait of a Lady* (already set up in NooJ), Charles Dickens' *Great Expectations*, Herman Melville's *Moby Dick*, as well as a contemporary oral corpus consisting of 25 transcribed *Larry King Live* programs from January 2000. Due to the limits of the initial PV grammar which incorporated a punctuation node <P>, however, NooJ failed to

© Springer International Publishing AG 2016
L. Barone et al. (Eds.): NooJ 2016, CCIS 667, pp. 169–181, 2016.
DOI: 10.1007/978-3-319-55002-2_15

recognize many occurrences of discontinuous PV. For example, while NooJ correctly identified the PV *sum up* in the following sentence, it did not recognize *think over*:

9. ... to feel the need of **thinking** things **over** and **summing** them **up**;

Since our overall goal is to automatically identify all PV in large corpora, [3] introduced a simplified PV grammar without the <P> constraint, as seen in Fig. 1. This grammar, along with two disambiguation grammars produced better recall, but overall accuracy dropped from 84% to 70–74%. In this article we will see how incorporating additional dictionaries and more sophisticated disambiguation grammars can produce better PV accuracy.

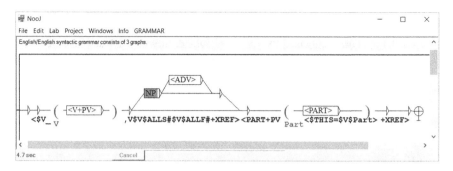

Fig. 1. NooJ PV grammar

2   Why so Much Noise?

2.1   Prepositions vs. Particles

One of the major difficulties in accurately identifying PV in English is the overlapping of particles and prepositions. Or as Talmy [4, p. 105] states, "a problem arises for English which, perhaps alone among Indo-European languages, has come to regularly position satellite and preposition next to each other in a sentence." At times, it is impossible to distinguish preposition from particle (or satellite) without a broader context or intonation. For example, the following sentence is ambiguous: the subject could be "looking above the fence" (preposition) or "examining the fence" (PV):

10. The neighbor **looked over** the broken fence.

However, if the PV is in the discontinuous format, then *over* is clearly a particle, with the only possible meaning of "examine":

11. The neighbor **looked** the broken fence **over**.

Furthermore, as Silberztein [5, pp. 270–271] illustrates, even in the discontinuous format, a sentence with a potential PV can also be ambiguous:

12. Max **switched** the computers **on** Monday.

This could either mean that "Max changed the location of the computers (switched) on Monday" (i.e., *on Monday* is a prepositional phrase) or that "Max started the computers on Monday" (i.e., *switch on* is a PV and the action occurred Monday). Since the difference could only be disambiguated through intonation or a larger context, NooJ keeps both interpretations in its Text Annotation Structure (TAS). However, there are many cases of prepositions, which have the same form as particles (e.g., *down, in, off, on*) that are not ambiguous, as in the following sentences, where *in* is clearly a preposition:

13. Do you remember what I **asked** you **in** Rome?
14. Osmond **asked in** a provokingly pointless tone.

When these verb + preposition instances are incorrectly annotated as PV, NooJ can remove the PV status automatically from the TAS with disambiguation grammars. But at the same time, NooJ must keep the TAS for accurate PV in similar phrases, such as:

15. Did you **ask** the prince **in** when he arrived?

2.2 Nouns vs. Verbs

Since NooJ does not use Part-of-Speech (POS) taggers, another problem encountered with the NooJ PV grammar was that many nouns were being misinterpreted as verbs (e.g., *a sudden break* vs. *break in* or *break up, the check* vs. *check in* or *check out, a loud cheer* vs. *cheer up, your figure* vs. *figure out, a couple of times* vs. *he times in* or *times out*, and so on). For example, NooJ would identify all of the following as potential PV:

16. standing there with his **hands** still **in** his pockets,
17. before Isabel noted a sudden **break in** her voice,
18. She **hands** her exam **in**.
19. He has tried to **break up** my relations with Isabel

Another disambiguation grammar is able to identify verbs that are obviously nouns (e.g., 16–17) by examining the preceding environment, and consequently removes the PV status for those, while keeping it for sentences 18–19. Although words such as *hand* and *time* may statistically appear more often as nouns than verbs, and *break* can also appear as both a noun and a verb, NooJ can still identify both *hand in* and *break up* when they appear as PV in large corpora without relying on POS taggers. We will now look at PV disambiguation grammars and supplementary dictionaries that can be added to NooJ which eliminate much of this noise, while not removing accurate PV.

3 Disambiguation Grammars

Each of the three disambiguation grammars specifies certain structures that are **not** to be assigned the TAS of <V+PV>. The specification <!V+PV> under the <V+PV> node in each grammar means "not a PV." Some grammars furthermore specify that the particle must be a preposition by listing <PREP> below the <PART> in the graph.

3.1 PV Disambiguation Grammar 1: Environment to Right of "PV"

The first disambiguation grammar (Fig. 2) considers the environment to the right of a supposed PV. This grammar is syntactically motivated, since in English, if the PV occurs with a pronoun object, it must generally be in the discontinuous format (e.g., *figure it out*, *look him up*, *take them away*). Thus if an object pronoun follows a supposed particle, it must be a preposition, as in the following:

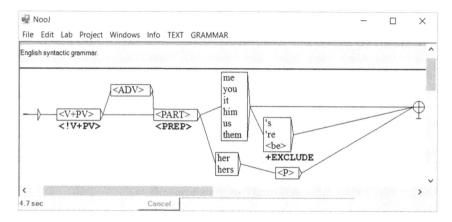

Fig. 2. PV disambiguation grammar 1

20. what sort of pressure is **put on** them back in Cuba
21. No, you won't believe what a comfort I **take in** it.
22. "Besides," said Mr. Pumblechook, **turning** sharp **on** me,

The pronoun *her*, since it is also the possessive adjective *her*, is not included with the other pronouns since it would produce too much silence (e.g., *locked up her apartment*, *pulling off her gloves*, *took up her parasol*, and so on). But if followed by punctuation, then we can be sure that it is not an adjective, as in the following, allowing NooJ to automatically remove the PV annotation from the TAS:

23. showing the same respectful interest in Isabel's affairs that Isabel was so good to **take in** hers.
24. rising and **bending over** her, as she rose from the bench.

Finally, using the NooJ functionality +EXCLUDE, this graph does not remove good PV, as in the following where the pronoun introduces another sentence (25) or is part of a *that* clause (with *that* deleted) followed by the verb *be* (26):

25. but from what I can **make out** you're not embarrassed
26. pundits earlier tonight **pointed out** it was Bradley doing poorly

The above sentences would be excluded from the automatic PV removal process, thus not creating silence.

3.2 PV Disambiguation Grammar 2: Environment to Left of "PV"

The second disambiguation grammar (Fig. 3) aims to identify verbs that are obviously nouns by examining the environment to the left of a supposed PV. Again, NooJ can automatically remove the PV status in the TAS if the preceding environment justifies it. A more detailed description of this grammar can be found in [3, pp. 157–162], but basically if a determiner or adjective appears immediately before the supposed PV, then the disambiguation grammar correctly assumes that it is a noun and removes the PV status from the TAS. This grammar successfully eliminates much noise derived from PV that overlap with nouns, such as *break in, check out, cheer up, figure out, hand in, head up, play out, sort out, take up, time in,* and so on.

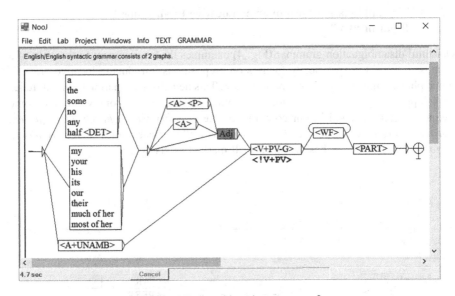

Fig. 3. PV disambiguation grammar 2

Furthermore, since certain adjectives can also be nouns, to avoid any silence, we limit this to only unambiguous adjectives using the NooJ functionality +UNAMB. In addition to determiners, this graph also includes introducers such as *couple of, great deal of, lot of, maximum amount of, much of her, most of her,* and other expressions that introduce nouns, but never verbs. This grammar then correctly removes the PV status from sentences 16–17 above (Sect. 2.2), while keeping it for examples 18–19. Due to sometimes overlapping of good PV and noise, this grammar does not specify that the particle must be a preposition. Again, the word *her* poses problems, since it can regularly be both an adjective and a pronoun. However, expressions such as *much of her time* and *most of her time* are adjectival, and thus this revised disambiguation grammar can identify sentences such as 27 but not remove the PV status in the TAS of correct phrasal verbs like 28:

27. she spends much of her **time in** thinking of him,
28. his traditions made her **push back** her skirts.

Finally, in this revised disambiguation grammar, we no longer indicate *o '*, since the word *o'clock* is now specified as an unambiguous adverb in the NooJ contractions dictionary:

29. o'clock, <o'clock,ADV>+UNAMB

This is similar to what we did with many compound nouns with dashes figuring in 19th century novels, such as *cash-box, watch-box, church-clocks, dish-cover, spring-time, summer-time, off-hand*, and so on, which also created a fair amount of initial noise with potential PV such as *box in, cover up, time in*, and *hand over*.

3.3 PV Disambiguation Grammar 3: Locative Environment to Right of "PV"

Our third disambiguation grammar (Fig. 4) examines the environment to the right of a supposed "PV," but specifically focuses on prepositions introducing locative prepositional phrases that are clearly not part of a PV. This new disambiguation grammar relies on a supplemental dictionary of locative nouns, NLoc. This dictionary contains some frequent locatives found in our corpora, such as *brewery yard, church, city, garden, library, sitting-room, street, yard*, and so on, as well as place names such as *America, London, Paris, Rome* – these nouns are all marked as N+Loc.

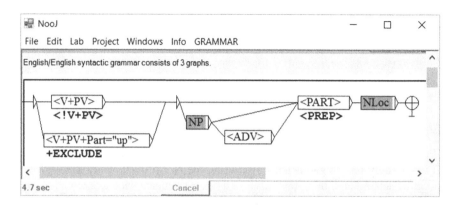

Fig. 4. PV disambiguation grammar 3

This new graph will eliminate the PV status in the TAS of many incorrectly identified phrasal verbs, mostly involving the prepositions *in* and *off*. For example the top part of the graph removes the PV TAS in continuous PV such as:

30. to be **asked in** church
31. as I had **done in** the brewery yard,

The bottom portion of the graph (i.e., the NP and ADV loop) does the same for discontinuous PV like:

32. Annie Climber was **asked** three times **in** Italy
33. what I **asked** you **in** Rome?
34. **take** these criminals **off** the street,
35. his exchange **took** place **in** the large decorated sitting-room

However, the +EXCLUDE path will not create silence by removing the status of genuine PV with the particle *up* such as:

36. He had been **brought up** in Paris,
37. took a tough guy to **clean up** New York
38. Touchett had better **shut up** her house
39. to **clean up** the streets

4   Auxiliary Dictionaries and Grammars

In addition to these three disambiguation grammars, we have also introduced auxiliary dictionaries and grammars that identify adverbial and adjectival expressions, as well as idioms whose prepositions (especially *in*, *off*, and *on*) can be mistaken for PV in our corpora, such as:

40. **asked** her **in a low tone** $\neq$ PV **ask in**
41. **put** the girl **on her guard** $\neq$ PV **put on**
42. **take an interest in** her $\neq$ PV **take in**

4.1   Adverbial and Adjectival Expressions

Under NooJ Syntactic Analysis, we now have two new graphs that identify a small subset of adverbial and adjectival expressions that have the potential to create a fair amount of noise when trying to accurately identify PV with the prepositions *in* and *on*. The Adverbials Grammar (Fig. 5) focuses in on expressions such as *at one time*, *in a low tone*, *in her lap*, *on one's mind*, and so on. When these expressions are detected by NooJ, they are marked as unambiguous adverbs (ADV+UNAMB), and thus neither the noun/verb (*time*), nor the preposition (*in*, *on*) can be associated with a PV. Thus the following are automatically eliminated from being potential PV (in bold), due to the unambiguous adverbial (underlined):

43. that you were <u>at one</u> **time in** intimate correspondence.
44. who had only **done** good **in** <u>the wrong way</u>.
45. she clasped her **hands** more tightly **in** <u>her lap</u>.
46. Touchett **has** been **on** <u>our minds</u> all winter

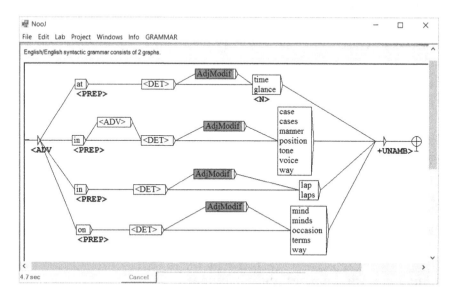

Fig. 5. Adverbials grammar

This graph can even identify more complex adverbials such as the following thanks to the AdjModif loop (Fig. 6):

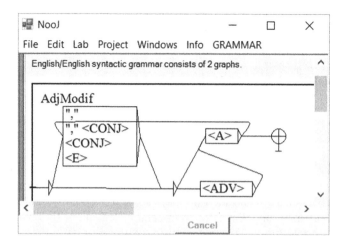

Fig. 6. AdjModif loop in adverbials grammar

47. **asked in** a provokingly pointless tone.
48. he then **asked in** a quick, full, slightly peremptory tone

The second additional grammar, Adjectival Expressions Graph (Fig. 7), identifies a few *be Prep C_1* idioms [6, 7] that also appear with certain support and operator verbs such as *have, put, take,* and so on, but have the potential to create noise when they are identified as PV (e.g., *have NP on, put NP on, take NP off*). Similar to the Adverbials Grammar, this graph labels certain expressions as unambiguous adjectives (A+UNAMB) and consequently eliminates PV noise from sentences such as:

49. she **had** been much **on** her guard
50. to **put** you **on** your guard?
51. what a fool he had been to **put** the girl **on** her guard
52. was waiting to **take** him **off** his guard.

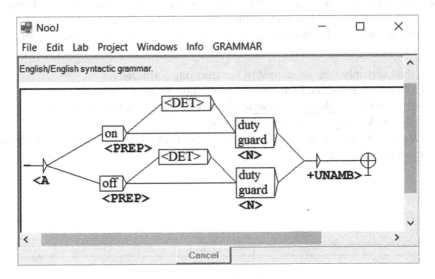

Fig. 7. Adjectival expressions graph

4.2 Idiom Dictionaries

Under NooJ Lexical Analysis, there is also a larger dictionary of simpler *Prep C_1* idioms that do not have multiple modifiers, such as *in character, in a haze, in the clouds, off duty, off the market, on sale, on a collision course, out of the question, over the top,* and so on. These are assigned the NooJ notation A+PrepC1+UNAMB, thus avoiding noise with potential PV using the particles *in, off, on, out, over*. For example, the expression *on pins and needles,* will not be confused with the PV *keep NP on* in the following sentence, since the potential particle *on* is part of an unambiguous *Prep C_1* idiom:

53. Your relations with him, while he was here, **kept** me **on** pins and needles;

Furthermore, there is dictionary that works in tandem with another grammar under NooJ Lexical Analysis that identifies a different type of idiom. These target the more complex $V\ C_1 Prep\ N_2$ idioms, such as *keep an eye on NP*, *take an interest in NP*, *take great pleasure in NP*, *take refuge in NP*, and so on, where the frozen noun can take a variety of determiners and modifiers. These are also called *CPN idiomatic expressions*, since the verb is followed by a frozen complement (C), then a preposition (P), and finally a free noun (N). The newly created NooJ CPN Dictionary contains entries such as:

54. take,V+Idiom+FXC+C1=interest+Prep=in+FLX=TAKE
55. take,V+Idiom+FXC+C1=comfort+Prep=in+FLX=TAKE
56. take,V+Idiom+FXC+C1=pleasure+Prep=in+FLX=TAKE
57. take,V+Idiom+FXC+C1=refuge+Prep=in+FLX=TAKE
58. take,V+Idiom+FXC+C1=pride+Prep=in+FLX=TAKE

The CPN grammar (Fig. 8) associated with this dictionary then recognizes these prepositional idioms in texts and marks them as unambiguous. Thus in the following sentences, NooJ will **not** recognize the PV *take NP in*, since the verb, noun, and preposition are clearly designated as being part of an unambiguous idiomatic expression:

59. I shall certainly take an interest in her marrying fortunately.
60. Gilbert Osmond should not have taken comfort in Miss Stackpole;
61. she would have taken a passionate pleasure in talking of Gilbert Osmond
62. Isabel took refuge in timorous vagueness.
63. Our heroine had always passed for a person of resources and had taken a certain pride in being one;

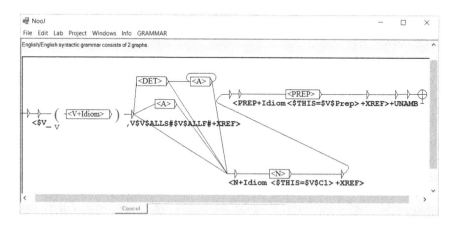

Fig. 8. CPN idiomatic expression grammar

5   Results and Future Work

5.1   Results

Due to the limits of our initial PV grammar [1, 2] which incorporated a punctuation node <P>, NooJ failed to find many occurrences of PV in large corpora. The revised PV grammar [3] allowed for a much better recall but created an inordinate amount of noise. However, with three disambiguation grammars, an adverbial expression grammar, and two idiom dictionaries and grammars, we have now succeeded in removing much of that noise. As we have seen, after dictionaries disambiguate certain word forms, annotate locatives, and identify idioms, supplemental grammars then examine the immediate left and right environments of potential PV and automatically eliminate the PV status in the TAS of false PV. They successfully identify verbs that are obviously nouns by verifying the preceding environment, as well as particles that are clearly prepositions in locative phrases or prepositional idiomatic phrases that follow would-be PV.

The results of the 2010 study [2] compared to today's are presented in Fig. 9. NooJ is now able to find almost twice as many correct discontinuous PV in the Henry James novel *Portrait of a Lady*. In addition, there are fewer problems of misidentification (e.g., nouns for verbs). Nevertheless, there are still major difficulties with prepositions following common verbs such as *do, have, take*. For example, NooJ cannot yet distinguish the accurate PV *have on* in sentence 64 from the noise in example 65:

64. She **had on** her hat and jacket.
65. She **had** a strange smile **on** her thin lips.

TEXT	Number of examples	Correct continuous examples	Correct discontinuous examples	Prepositions	Misidentifications	Percentage of incorrect	Percentage of correctly identified phrasal verbs
Portrait of a Lady (2010)	583	405	83	44	51	16.30%	83.70%
Portrait of a Lady (2016)	658	427	152	61	18	12.01%	87.99%

Fig. 9. Comparison of PV identification in *Portrait of a Lady*: 2010 study vs. today

Still, recall of discontinuous PV has improved and accuracy has gone from 84% to 88%. These problems will eventually be resolved when more local grammars are built for English and NooJ's syntactic parser is able to recognize entire English sentences. In fact, as Silberztein [8, p. 188] has shown for French, NooJ's syntactic parser "allows linguists to build and accumulate larger syntactic grammars up to the sentence level." This could help us further fine-tune the results of our local disambiguation grammars.

5.2   Future Work

Since Kennedy's [9] classic study, phrasal verbs have often been classified as pleonastic or colloquial variants of simple verbs (*finish up* vs. *finish*; *cough up* vs. *pay*), which have even been "blamed" on an American influence. Thim [10] challenges many of Kennedy's assumptions. Our initial work on *Portrait of a Lady* is very promising and we hope to shed more light on the history of phrasal verbs by automatically annotating them in the works of Dickens and Melville. In fact, in preliminary analyses, we have been able to establish that the 19th century British writer Charles Dickens tends to use more phrasal verbs than his American counterpart, Herman Melville, as shown in Fig. 10.

TEXT	Year	Word forms	Phrasal Verbs	Phrasal Verbs per 1,000 words
Moby Dick	1851	218,390	794	3.64
Great Expectations	1867	188,948	1040	5.50
Larry King Live	2000	228,950	803	3.51

Fig. 10. Comparison of PV usage per 1,000 words of text

For example, there are only 3.6 PV per 1,000 words in the American work *Moby Dick*, published in 1851, but 5.5 PV per 1,000 words in the British work *Great Expectations*, written in 1867. Furthermore, there are even fewer PV in our oral corpus: 3.5 PV per 1,000 words in the *Larry King Live* interview transcripts. Although we have not yet achieved total accuracy and recall in recognizing phrasal verbs in large corpora, Lexicon-Grammar tables and the NooJ linguistic development environment, including disambiguation grammars, are powerful tools that can help solve problems in Natural Language Processing, as well as resolve discrepancies in Historical Linguistics.

References

1. Machonis, P.A.: NooJ: a practical method for parsing phrasal verbs. In: Blanco, X., Silberztein, M. (eds.) Proceedings of the 2007 International NooJ Conference, pp. 149–161. Cambridge Scholars Publishing, Newcastle upon Tyne (2008)
2. Machonis, P.A.: English phrasal verbs: from lexicon-grammar to natural language processing. South. J. Linguist. **34**(1), 21–48 (2010)
3. Machonis, P.A.: *Sorting* NooJ *out* to *take* multiword expressions *into account*. In: Vučković, K., Bekavac, B., Silberztein, M. (eds.) Automatic Processing of Various Levels of Linguistic Phenomena: Selected Papers from the NooJ 2011 International Conference, pp. 152–165. Cambridge Scholars Publishing, Newcastle upon Tyne (2012)
4. Talmy, L.: Lexicalization patterns: semantic structure. In: Shopen, T. (ed.) Lexical Forms in Language Typology and Syntactic Description, pp. 57–149. Cambridge University Press, New York (1985)

5. Silberztein, M.: Formalizing Natural Languages: The NooJ Approach. Wiley ISTE, London (2016)
6. Machonis, P.A.: Support verbs: an analysis of *be prep X* idioms. SECOL Rev. **12**(2), 95–125 (1988)
7. Giannasi, R.: Expressions figées: *be Prep X* en anglais américain. M.A. thesis. Mémoires du CERIL 7, pp. 117–202. Université Paris 7, Paris (1990)
8. Silberztein, M.: Syntactic parsing with NooJ. In: Ben Hamadou, A., Mesfar, S., Silberztein, M. (eds.) NooJ 2009 International Conference and Workshop on Finite State Language Engineering, pp. 177–189. Centre de Publication Universitaire, Sfax (2010)
9. Kennedy, A.G.: The Modern English Verb-Adverb Combination. Stanford University Press, Stanford (1920)
10. Thim, S.: The English Verb-Particle Construction and its History. De Gruyter Mouton, Berlin (2012)

NooJ Local Grammars for Endophora Resolution

Mario Monteleone[✉]

Dipartimento di Scienze Politiche, Sociali e della Comunicazione,
Università degli Studi di Salerno, Salerno, Italy
mmonteleone@unisa.it

Abstract. In linguistics, the term endophora [1] defines the co-reference of an expression with another expression either before or after it. The endophoric relationship is often spoken of as one expression 'referring to' another. Endophora mainly takes places when, in a same sentence, a pronoun and a (proper or common) noun co-occur referring (back or forward) to one another. This mechanism is called endophoric deixis. The most used and known kinds of endophora are anaphora (the use of an expression which depends specifically upon an antecedent expression), and cataphora (the use of an expression which depends upon a postcedent expression). The anaphoric term is called an anaphor (a word, such as a pronoun, used to avoid repetition; the referent of an anaphor is determined by its antecedent), while the cataphoric term is called a cataphor (a word that refers to or stands for another word used later).

Keywords: NooJ · NooJ local grammars · Lexicon-Grammar · Endophoric deixis · Endophora resolution · Natural language processing · Rule-Based parsing · Anaphora disambiguation · Cataphora disambiguation

1 Introduction

In linguistics, the term endophora [1] defines the co-reference of an expression with another expression either before or after it. One expression provides the information necessary to interpret the other. The endophoric relationship is often spoken of as one expression 'referring to' another. The most used and known kinds of endophora are anaphora [1] (the use of an expression which depends specifically upon an antecedent expression), and cataphora [1] (the use of an expression which depends upon a postcedent expression). The anaphoric term is called an anaphor (a word, such as a pronoun, used to avoid repetition; the referent of an anaphor is determined by its antecedent). On the contrary, the cataphoric term is called a cataphor (a word that refers to or stands for another word used later). Endophora mainly takes places when, in a same sentence, a pronoun and a (proper or common) noun co-occurs referring (back or forward) to one another. This mechanism is called endophoric deixis (ED). For example, in the sentence

© Springer International Publishing AG 2016
L. Barone et al. (Eds.): NooJ 2016, CCIS 667, pp. 182–195, 2016.
DOI: 10.1007/978-3-319-55002-2_16

"**Sally** arrived, but nobody saw **her**", the pronoun **her** is an anaphor[1], referring back to the antecedent **Sally**. In the sentence "Before **her** arrival, nobody saw **Sally**", the pronoun **her** refers forward to the postcedent **Sally**, so **her** is now a cataphor. Also in Italian, endophora may address sentence semantics and is governed by co-occurrence, restriction selection and inalienability rules [2]. As such, endophora can be formally described by means of NooJ local grammars. Therefore, the main topics of this paper will be the formalization of the morpho-syntactic mechanisms that govern endophora and ED, and their conversion into NooJ local grammars instructions [3]. During this procedure, we will give particular attention to some Italian language peculiarities that may function as obstacles to a correct Endophora Resolution (ER). These peculiarities are:

1. With regard to lexicon, the presence of a high categorial ambiguity of pronouns which very often share the same form and inflection features with some Italian determiners and adjectives, an aspect which may create structural and semantic ambiguity;
2. With regard to syntax, and mainly in simple-sentence paraphrases, the fact that the deixis can also be realized in a disconnected and discontinuous way, that is between elements which in a sentence or discourse[2] are physically distant from each other.

For instance, let us consider the Italian discourse *La mela non la mangia tutta John; la metà la mangia Paolo.* (John does not eat the whole apple; one half of it is eaten by Paul). An Italian native speaker would not have particular problems in correctly tagging the four different occurrences of "*la*", concluding that "*la mela*" and "*la metà*" are sequences of the kind "determiner+noun", while the two "*la mangia*" are sequences of the kind "pronoun+verb". However, and although it may seem unnecessary to emphasize it, we must remember that the parsing performed by humans is a complex mechanism, which allows rapid and continuous movements forward and backward inside linguistic expressions brought to our attention. In short, human parsing puts linguistic competence to work on various types of utterances, which we understand and disambiguate thanks to the morpho-syntactic knowledge we have about our or other languages. But [3–6] well state that in Natural Language Processing (NLP) the formal reproduction of human parsing mechanisms requires a taxonomic descriptive approach, as regards both lexicon and morpho-syntax of each and any language. Besides, as we will see in the following pages, to achieve a correct ED parsing, it is necessary to create precise formal grammars, functional in providing reliable and reusable results. This means that we will need to study any occurrence contexts before creating formal grammars in which to disambiguate the Italian ED forms previously exposed. Such a taxonomic study of

---

[1] A proper identification of these and similar anaphoric and/or cataphoric referrals must be performed according to sentence semantics, avoiding any possible form of interpretation. This implies the fact of attesting the co-reference between **her** and Sally, and that **her** do not refer to another esophoric proper noun. In the following pages, we will return to this issue more broadly.

[2] As it is defined in [9, page 173], discourses are utterances formed by two or more simple sentences, connected by means of either paratactic or hypotactic conjunctions.

occurrence contexts will force us to enlarge our analysis, including not only such forms but also other linguistic items or data. We will see in detail which complex parsing rules are necessary to disambiguate Italian deictic pronouns from determiners and adjectives. We will call Endophora Resolution (ER) this procedure of disambiguation. To achieve it, we will adopt Lexicon-Grammar theoretical and practical framework. After creating a sufficiently exhaustive corpus, we will build a set of local grammars with NooJ [3], apply them to our corpus, and analyze the results of this application.

2   Different Types of Endophora

Depending on the syntactic context in which they occur, we may have two different types of Endophora, namely Simple-Sentence Endophora (SSE) and Discourse Endophora (DE). In SSE, ED are governed by the syntactic behavior of a given semantic predicate [4–6] occurring inside a simple sentence as it is defined by Lexicon-Grammar [7, 8]. In such cases, we may achieve ED parsing formalizing the co-occurrence, selection-restriction and transformational rules governed by that specific semantic predicate. As for this type of ED parsing, pronominalization is the most important transformational rule to take into account.

On the contrary, even if it is governed by the same rules as SSE, DE occurring within discourses may present two main problems. First, the fact that the sentences forming a discourse may not be in the SVO active form and/or in the present tense; this would force parsing to cope also with transformations and compound verb forms. Second, the fact that DE may correlate text items located even quite far from one another. As we will see, with reference to formal grammars for SSE parsing, in such cases ED reconstruction will require the use of more complex procedures.

2.1   Italian Morpho-Syntactic Obstacles to ER

As already stated, Italian morpho-syntax presents two main obstacles to correct ER, the first one arising from Lexical Ambiguity (LA), the second one from verb inflection and particular syntactic predicate behaviors, as for instance Pro-Drop.

Italian Pronouns and (pre)Determiners. As for Italian pronouns and (pre)-determiners[3], LA derives from their being often homographs, also sharing the same inflectional features. We list hereby the 63 Italian words which in all their singular, plural, masculine and feminine forms, may contemporarily function as pronouns or pre-determiners:

---

[3] Pre-determiners are modifiers of other determiners, nouns or articles; they can be classified into three main groups, such as multipliers, fractions, and intensifiers. They state precisely or suggest approximately the amount or the number of a noun. An intensifier is also called a booster. Despite having little meaning in itself, it adds, however, force to the meaning of another word or a phrase it modifies. We can group pre-determiners and modifiers by the noun types they quantify.

alcuno, altrettanto, altro, altrui, ambedue, antipenultimo, assai, certi, ciascheduno, ciascuno, dimolti, diversi, due, entrambi, il, la, lo, loro, medesimo, mio, moltissimo, molto, nessuno, niente, niuno, nostro, nulla, nullo, ognuno, parecchio, penultimo, poco, primo, propio, proprio, prossimo, qualcheduno, qualcosa, qualche, qualcuno, quale, qualunque, quanto, quartultimo, quel, quello, questo, quintultimo, sestultimo, stesso, suo, talaltro, tale, talun, taluno, tantino, tanto, terzultimo, tot, troppo, tuo, tutto, ultimo.[4]

Due to the LA of these words, we can precisely identify their correct morphosyntactic functions and roles only parsing the simple sentences in which they occur. For instance, let us consider the two following Italian examples, in which we find (in brackets) words normally lemmatized as both pronouns and pre-determiners:

*Ho visto (**molte+tante+diverse+alcune+poche+svariate**) case*
(I saw (many+several+various+some+few+different) houses)

*Di case, ne ho viste (**molte+tante+diverse+alcune+poche+svariate**)*
(As for houses, I saw (many+several+various+some+few+different) of them)

As for the first example, we observe that the words in brackets quantify to a certain extent the noun they precede ("case"), not creating with it any kind of ED. Therefore, we can state that such words occur inside a syntactic pattern of the kind pre-determiner +noun (PREDET+N). On the contrary, as for the second example, we observe that all the words into brackets are in ED with the name preceding them (once again "case").Therefore, we can state that they occur inside a syntactic pattern of the kind verb +pronoun (V+PRON). An additional consideration is that V+PRON patterns are always present inside simple sentences subjected to extrapolations (i.e. leftward dislocations).

These two different syntactic patterns are highly recurring inside Italian texts. Therefore, as for LA disambiguation and ED reconstruction, they are the most appropriate to study in order to formalize an efficient and effective parsing method. Besides, Italian pronouns and (pre)determiners may form other syntactic patterns, which we list hereby, with words belonging to different parts of speech (POS). It is worth stressing that in all these patterns, Italian pronouns and (pre)determiners are in contrastive distribution, which means that an acceptable sequence including a pronoun becomes unacceptable if we substitute such a pronoun with a (pre)determiner:

- *PRON+ADJ (pronoun+adjective)
- *PRON+N (pronoun+noun)
- PRON+PREP (pronoun+preposition)
- PRON+V (pronun+verb)
- PRON+ADV (pronoun+adverb)

---

[4] These words may be approximately translated into English with "some, equally, other, others, both, antepenult, rather, some, each one, each, many, several, two, both, the/it/her, the/it/his, them, the same, my, much, much, nobody, nothing, no one, ours, nothing, nil, each one, a lot, next to last, little, first, just/own, just/own, next, somebody, something, somewhere, someone, who/whom, whatever, what, fourth last, that, that, this, fifth last, sixth last, himself, his, someone else, such, someone, anybody, little, much, third last, tot, too, you, all, last". At any rate, many of these translations could be faulty in specific sentence contexts.

- PRON+(E+DET)+PRON (pronoun+a possible determiner+pronoun[5])
- (COMPUND) PREP+PRON+DET+N (compound or not compound preposition +pronoun+determiner+noun)
- (E+DET)+PRON +(,+.+;+:) (a possible determiner+pronoun+any diacritic)
- DET+'+PRON+(,+.+;+:) (apostrophized determiner+pronoun+any diacritic)

In the following pages, in order to build our parsing method, we will formalize such patterns inside a set of NooJ Local Grammars.

Pro-Drop effects on ED in Italian. With regard to verb inflection and syntactic features, it is widely known that together with all most spoken Romance languages but French, Italian is a Pro-Drop language, especially as regards the dropping of subject personal pronouns. Due to this feature, Pro-Drop languages normally use verb morphology and conjugation to express the morphological number of the pronouns they drop.[6]On the contrary, Romance languages cannot use verb inflection to specify the gender of the dropped[7] pronoun. As we will see in the following Italian passage, pronoun dropping actually creates a different kind of deixis, which does not pertain to syntax, but to text interpretation:

> **Vide** un uomo uscire dalla casa. **Era** di spalle, e perciò non **fu** in grado di riconoscerlo, guardandolo in viso. Di colpo, l'uomo cominciò a correre. Lo **vide** scomparire fra la folla, poi riemergere, e voltare dietro l'angolo del primo isolato. ((***He+She***) *saw a man coming out from the house, his back turned, so (**he+she**) couldn't look him in the face and recognize him. Suddenly, the man started to run.* ((***He+She***) *saw him vanishing in the crowd, then coming out again, finally turning around the corner of the first block.*)

The English translation of this passage may help in better underlining the problems Pro-Drop mechanisms may cause to correct ED and ER. Actually, when a personal pronoun drops, especially before a verb at the third person singular or plural, we have no mean to reconstruct ED and its mechanisms. Also, we cannot detect the gender of the pronouns dropped. In this sense, pronoun dropping in Italian produce exophora, or better Exophoric Reference (ExR) [10], which is a phenomenon pertaining more to

---

[5] These patterns are rather rare, and concern word sequences as "alcuni I quali, alcuni che" (some who/whom).

[6] Some minimal examples of Pro-Drop sentences in Romance languages are:
Italian: *(E + Io) Mangio una pizza / (E + Tu) Mangi una pizza*
French: ***Je*** *mange une pizza /* ***Tu*** *manges une pizza*
Spanish: *(E + Yo) Como una pizza / (E + Tú) Comes una pizza*
Portuguese: *(E + Eu) Como uma pizza / (E + Tu) Comes uma pizza*
Romanian: *(E + Eu) Mânca o pizza / (E + Tu) Mănânci o pizza.*

[7] For instance in:
Italian: *(E +* ***Lui*** *+* ***Lei****) Mangia una pizza*
French: (***Il*** *+* ***Elle***) *Mange une pizza*
Spanish: *(E +* ***Él*** *+* ***Ella***) *Come una pizza*
Portuguese: *(E +* ***Ele*** *+* ***Ela****) Come uma pizza*
Romanian: *(E +* ***El*** *+* ***Ea****) Mănâncă o pizza.*

pragmatics than to morpho-syntax, being it extralinguistic and constrained by co-text[8] occurrences (words and/or propositions). ExR leads us outside the text we are reading, forcing us to "interpret" rather than analyze. Considering that the formalization of inter-pretation mechanisms is a task that falls beyond those of our paper, in the following paragraphs we will show how to build and apply only NooJ local grammars for ER, and we will not deal with ExR.

3   NooJ Local Grammars

The correct achievement of ER required the location and formalization of the word contexts in which the previous 63 Italian words characterized by LA may occur. To locate such contexts, we built concordances applying the NooJ grammar of Fig. 1 to our corpus[9]. This passage produced 17711 occurrences (concordances) out of 159711 tokens, including 124497 word forms, 20 digits, and 35194 delimiters:

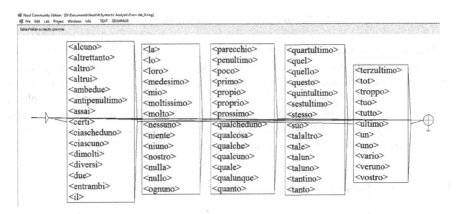

Fig. 1. Pron_Det_0.nog grammar for concordance building (This grammar presents four different nodes only due to typographic reasons: putting all the words inside a single node would have made it impossible to have a readable representation of the whole grammar.)

---

[8] In [11, page 46], co-text is defined as follows: "In our discussion so far we have concentrated particularly on the physical context in which single utterances are embedded and we have paid rather little attention to the *previous discourse* co-ordinate. Lewis introduced this co-ordinate to take account of sentences which include specific reference to what has been mentioned before as in phrases like *the aforementioned*. It is, however, the case that any sentence other than the first in a fragment of discourse, will have the whole of its interpretation forcibly constrained by the preceding text, not just those phrases which obviously and specifically refer to the preceding text, like *the aforementioned*. Just as the interpretation of the token q in the child's representation of 'without to disturb the lion' and the token [p] in [greipbritn] are deter-mined by the context in which they appear, so the words which occur in discourse are constrained by what, following Halliday, we shall call their co-text".

[9] Our corpus files is a novel and contains 170 pages, ie 122.315 words.

Figure 2 shows an excerpt of the concordances built applying this grammar to our corpus. The left and right contexts of concordances were set at 45 items, to include and analyze a relevant number of co-occurring words:

Fig. 2. An excerpt of the concordances built

The analysis of these concordances brought us to build different types of ER grammars which we will discuss in the following paragraphs. Yet, some preliminary observations are possible: anaphora is more recurrent than cataphora; anaphoric elements seem to occur not very separate one from the other; the formalization of endophora constraints seems possible using NooJ variables, and pushing the analysis forward and backward inside simple sentences and discourses; the main morph-syntactic constraint to cope with is agreement (in gender and number).

4   The Grammars

As already stated, the building of interpretation grammars falls beyond the compasses of this paper. For instance, our corpus includes the following discourse:

> *E dopo che tutto fu finito,* (**pro-drop**) *rimasero a guardare, alla poca luce della luna. "Una pioggia, e non si vedrà più niente", disse l'uomo. Gli **altri** non dissero nulla.* (Once it was all over, **they** went on watching, in the dim light of the moon. "Let it rain and nothing will be visible anymore," the man said. The **others** said nothing.)

As for the pronouns it includes, we will only be able to detect their gender and number; on the contrary, we will not be able to reconstruct their deictic mechanisms, considering that their referents are to be found outside the discourse itself, somewhere in the text, even in a not so close part.

4.1 The Necessity to Analyze Samples of ED

Lexicon-Grammar has proven that natural language is a combinatory of elements [12] governed by the syntactic-semantic features of predicates. In this regard, the possible combinations of the words of a language must be calculated in exponential terms. This implies that a fine-grained report of all ED possible contexts needs a very large processing, and a very high number of written pages. Since we do not have such material premises, in the next few paragraphs we will focus on the delineation of a possible ER method, analyzing some LAs and building a grammar for ED management. Therefore, the Italian words upon which we will focus our attention are *altro* (other) and *la* (the, her) that can be used as both pronoun and (pre)determiners.

4.2 Disambiguation of the Word *altro*

Figure 3 shows the first grammar built to disambiguate the word *altro* as a pronoun or a pre-determiner:

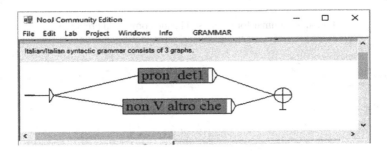

Fig. 3. Endophora_02.nog

This non-deterministic finite state automaton [3] embeds two more grammars (in yellow), the content of which is shown below. To test their proper functioning, we used the debug option [13] of NooJ grammars, which colors in greens the propositions perfectly matching with the instructions formalized. As for Figs. 4 and 5, such option confirms that the grammars built correctly tag the word *altro* when it is used inside compound pronouns as *qualcun altro che* (some other who/whom) and inside semi-fixed propositions as *non V altro che V* (do not V anything but V):

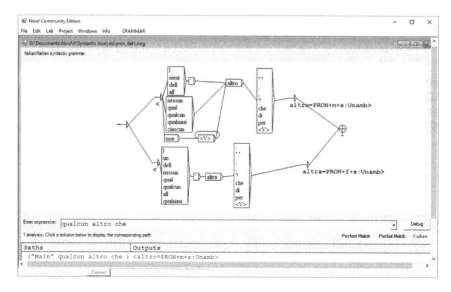

Fig. 4. Grammar for *altro* used inside compound pronouns

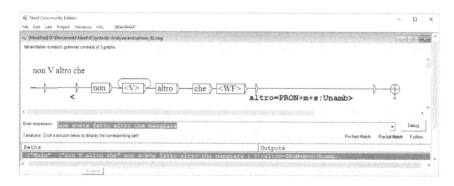

Fig. 5. Grammar for *altro* used inside semi-fixed propositions

The grammar in Fig. 6 is used to locate all the 63 Italian words characterized by LA when they are used as (pre)determiners, that is to say when they are not used as pronouns ((PRE) DET=:NO_PRON tag). The sample phrase analyzed is *altri giovani francesi* (other French young people):

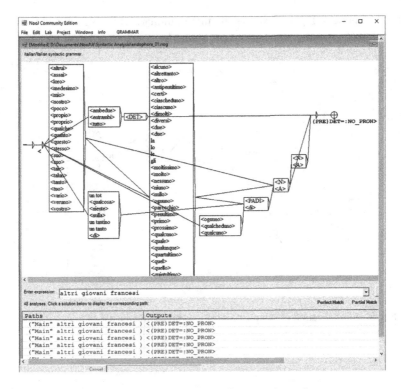

Fig. 6. Grammar for *altro* used as a predeterminer

4.3   Resolving LA and ED for the Word *la*. Some Samples

Concerning the word *la*, to make some examples of LA resolutions and ED reconstruction, we built four different local grammars, two of which were embedded in the main

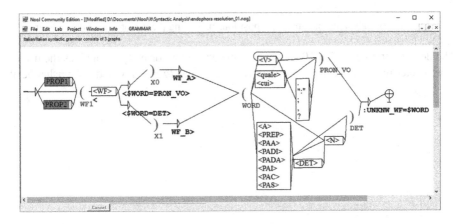

Fig. 7. LA disambiguation for the word *la*

ones (yellow nodes). Figure 7 shows the first grammar, while Figs. 8 and 9 show the formalizations of specific propositions preceding *la* in its two different uses:

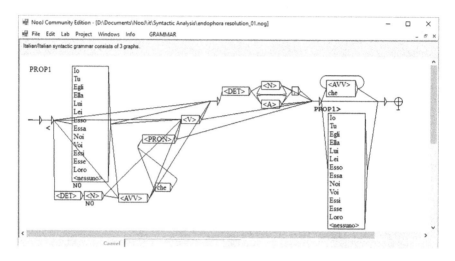

Fig. 8. Formalization of sample propositions preceding the use of *la* as a pronoun

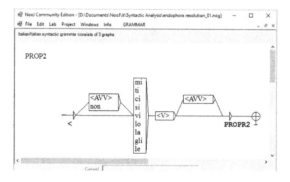

Fig. 9. Formalization of sample propositions preceding the use of *la* as a determiner

The debug window of Fig. 10 shows how the grammar correctly recognizes the use of *la* as a determiner inside the simple sentence *La spinge la bontà* (Goodness pusher her):

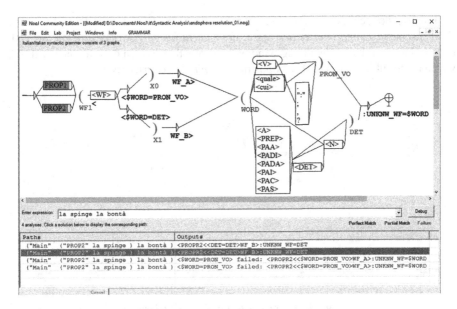

Fig. 10. WF=DET (*la*=:determiner)

As well, the debug window of Fig. 11 shows how this same grammar correctly recognizes the use of *la* as a pronoun, in the same simple sentence, but with a reversed phrase order, i.e. *La bontà la spinge*:

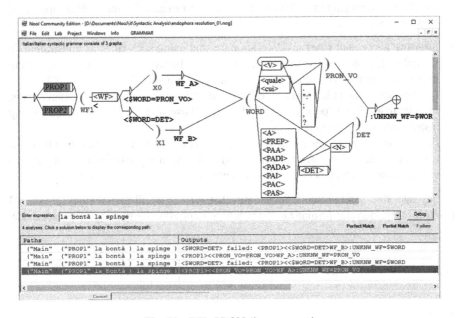

Fig. 11. WF=PRON (*la*=:pronoun)

Finally, the local grammar of Fig. 12 reconstructs the ED between *lei* and *le* in the short discourse *Prima che lei arrivasse, nessuno la aveva vista* (Before she arrived, nobody had seen her). The debug window confirms that *la* is in co-reference with *lei*, which is the subject (N0) of the proposition *Prima che lei arrivasse*:

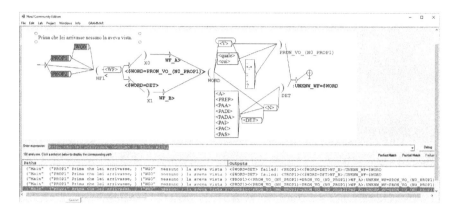

Fig. 12. ED reconstruction between *lei* and *la*

5   Conclusions and Further Steps

The quantity of formal instructions used to resolve LA and reconstruct ED shows us that ER is a very challenging task. Thanks to NooJ and Lexicon-Grammar NLP approach, we will improve our grammars:

- Inserting agreement rules inside grammars, and using such rules as constraints on variables, thus allowing better ED reconstructions between, for instance, co-referring pronouns or nouns having the same number and/or gender;
- Labeling electronic dictionary entries with morpho-grammatical and semantic tags, in order to evidence specific POS LA[10]; we will use such tags as abstracted parsing instructions inside grammars.
- Formalizing co-occurrence and selection-restriction rules of specific semantic predicate sets to reach a fine-grained syntactic description, and predict all possible acceptable pronominalizations inside simple sentences.

At any rate, this rapid sketch demonstrates that *even* in a still partial study phase, ED based on morpho-syntax formalization tends to be more effective than statistics-based one.

---

[10] For instance, adjectives, nouns, verb (past participles), compound pre-determiners, and pronouns, ie POS elements with a high LA potential level.

References

1. Crystal, D.: A Dictionary of Linguistics and Phonetics, 2nd edn. Basil Blackwell, New York (1985)
2. Elia, A., Martinelli, M., D'Agostino, E.: Lessico e Strutture Sintattiche. Liguori, Napoli (1981)
3. Silberztein, M.: La formalisation des langues: l'approche de NooJ. ISTE Editions, Londres (2015)
4. Silberztein, M.: Analyse et génération transformationnelle avec NooJ. In: Elia, A., Iacobini, C., Voghera, M. (eds.) Proceedings of the 47th Annual Meeting of the Italian Linguistic Society, Livelli di Analisi e Fenomeni di Interfaccia, pp. 225–242. Bulzoni, Rome (2015)
5. Silberztein, M.: NooJ V4. In: Koeva, S., Mesfar, S., Silberztein, M. (eds.) Formalising Natural Languages with NooJ 2013. Selected Papers from the NooJ 2013 International Conference, pp. 1–12. Cambridge Scholars Publishing, Newcastle (2014)
6. Silberztein, M.: NooJ Computational devices. In: Donabédian, A., Khurshudian, V., Silberztein, M. (eds.) Formalising Natural Languages with NooJ, pp. 1–13. Cambridge Scholars Publishing, Newcastle (2013)
7. Gross, M.: Les bases empiriques de la notion de prédicat sémantique. Langages (63), 7–52 (1981)
8. Elia, A., Vietri, S., Postiglione, A., Monteleone, M., Marano, F.: Data mining modular software system. In: Arabnia, H.R. Marsh, A., Solo, A.M.G. (eds.) Proceedings of the 2010 International Conference on Semantic Web & Web Services, SWWS 2010, pp. 127–133. CSREA Press, Las Vegas (2010)
9. Elia, A., Postiglione, A., Vietri, S., De Bueriis, G., Monteleone, M., Marano, F.: Semantics from lexis grammar. In: Vitas D., Krstev C. (eds.) Proceedings of the 29th International Conference on Lexis and Grammar, pp. 121-130. Faculty of Mathematics, University of Belgrade, Belgrade (2010)
10. Gross, M. Grammaire transformationnelle du français: Tome 2, Syntaxe du nom. Cantilène, Paris (1986)
11. Boons, J.-P., Guillet, A., Leclère, C.: La structure des phrases simples en français. Droz, Genève (1976)
12. Harris, Z.S.: Papers in Structural and Transformational Linguistics. Reidel Publishing Company, Dordrecht (1970)
13. Halliday, M.A.K.: An Introduction to Functional Grammar, 3rd edn. Oxford University Press Inc., New York (2004)

Paraphrases for the Italian Communication Predicates

Alberto Maria Langella[✉]

Department of Political, Social and Communication Sciences,
University of Salerno, 84084 Fisciano, Italy
allangella@unisa.it

Abstract. In this paper, I will discuss the production of paraphrases for the Italian communication predicates. These predicates represent, statistically, a significant part of the Italian lexicon-grammar class 47. My theoretical assumption is the Harrisian concept of transformation, which is an equivalence relation between sentences with a change in the grammatical form while preserving the morphemes and the grammatical relations between them. We deal with paraphrases like pronominalization of the arguments, reduction to zero of the arguments, permutations of the arguments and passivization. Each of the previous types of paraphrases can be combined with each other according to certain syntactic constraints.

Keywords: Paraphrase · Syntactic constraints · Support verb · Morphism

1 Introduction

This article is my second work on the production of paraphrases with NooJ, and comes after my study of the Italian paraphrases of the psychological predicates. My theoretical assumption is the Zellig Sabbettai Harris concept of paraphrase, on which he began to work in the early 50 s. Harris probably borrowed it from category theory[1], at that time a pioneering and new branch of mathematics developed in the early 40 s by the American mathematician Saunders Mac Lane and the Polish-born American mathematician Samuel Eilenberg. In category theory, mathematicians use the so-called morphisms, which are structure-preserving maps from one mathematical object to another. Harrisian paraphrases may be seen as a crypto-morphism: two mathematical objects are crypto-morphic if they are equivalent but not obviously equivalent. In fact, according to Harris a sentence is a paraphrase of another sentence if a change occurs in the morphophonemic shape of the transformed sentence while preserving the original lexical morphemes and meaning (e.g. passivization).

Harris' pioneering research on information retrieval (IR) dates back to the late 50 s and today many researchers in the information retrieval community are starting to

---

[1] For an introduction to category theory to see Mac Lane, S., Eilenberg, S.: General Theory of Natural Transformations. In: Transactions of the American Mathematical Society, Vol. 58, No. 2, pp. 231–294. American Mathematical Society Press, New York (2001).

© Springer International Publishing AG 2016
L. Barone et al. (Eds.): NooJ 2016, CCIS 667, pp. 196–207, 2016.
DOI: 10.1007/978-3-319-55002-2_17

emphasize the importance of predicate-argument relation and of paraphrases in order to improve and make our language processing capability more accurate:

"Most conventional approaches to information retrieval (IR) deal with words as independent terms. In query sentences and documents, however, dependencies exist between words. To capture these dependencies, some extended IR models have been proposed in the last decade (Jones, 1999; Lee et al., 2006; Song et al., 2008; Shinzato et al., 2008). These models, however, did not achieve consistent significant improvements over models based on independent words.

One of the linguistic reasons for this is the linguistic variations of syntax, that is, languages are syntactically expressed in various ways. For instance, the same or similar meaning can be expressed using the passive and the active voice in a sentence. Previous approaches based on dependencies cannot identify such variations. This is because they use the output of a dependency parser, which generates syntactic (grammatical) dependencies built upon surface word sequences."

In Sect. 2, I provide a brief description of the communication predicates belonging to the Italian lexicon-grammar class 47. In Sect. 3, I describe the NooJ syntactic grammar devised for the production of paraphrases and the electronic dictionary on which the syntactic grammar relies.

2   The Italian Communication Predicates

The communication predicates we have taken into account are a subset of the Italian lexicon-grammar[2] class 47. Their definitional structure is N0 V ChF1 a N2. N0 is in most cases and with very few exceptions a human noun (N+Hum). N1 is a clause to which we can assign the lexicon-syntactic role of "message". The verbs of this class are *annunciare* (*to announce*), *comunicare* (*to communicate*), *confermare* (*to confirm*) and so on, in sentences like:

(1) *Mike annuncia la sua partenza a John*
      (*Mike announces his departure to John*)
(2) *Mike comunica la sua partenza a John*
      (*Mike communicates his departure to John*)
(3) *Mike conferma la sua partenza a John*

Nominalizations are possible. *Fare* (*to make*) and *dare* (*to give*) are largely predominant as "carrier verbs"[3]. In most cases, the verbs of this class can accept simple nouns as a direct object. But according to Harrisian grammatical principles these nouns must

---

[2] For an introduction to the Italian lexicon-grammar I refer the reader to Elia, A., Martinelli, M., D'Agostino, E.: Lessico e Strutture della Sintassi. LoffredoEditore, Napoli (1981).

[3] The expression "carrier verbs" has been introduced in linguistics by Zellig Harris. It refers to verbs which behave syntactically like the traditional copula. Before Harris, Otto Jespersen used the expression "light verbs" in order to refer to the same concept.

be interpreted as reductions[4] of clauses. These verbs can be interpreted as "communication operators" with a transfer of information from N0 to N2. They accept nominalizations like the following:

(4) *Mike fa l'annuncio della sua partenza a John*
     (*Mike makes the announcement of his departure to John*)
(5) *Mike dà comunicazione della sua partenza a John*
     (*Mike gives notice of his departure to John*)
(6) *Mike dà conferma della sua partenza a John*
     (*Mike gives confirmation of his departure to John*)

The N1s in sentences number (1), (2) and (3) are in distributional equivalence with *that*-clauses:

(7) *Mike annuncia* (E+*il fatto*) *che partirà a John*
     (*Mike announces* (E+*the fact*) *that he will leave to John*)
(8) *Mike comunica* (E+*il fatto*) *che partirà a John*
     (*Mike communicates* (E+*the fact*) *that he will leave to John*)
(9) *Mike conferma* (E+*il fatto*) *che partirà a John*
     (*Mike confirms* (E+*the fact*) *that he will leave to John*)

The N1s can even be in distributional equivalence with infinitive clauses:

(10) *Mike annuncia di partire a John*
      (*Mike announces his leaving to John*)
(11) *Mike comunica di partire a John*
      (*Mike communicates his leaving to John*)
(12) *Mike conferma di partire a John*
      (*Mike confirms his leaving to John*)

In the previous sentences, *Mike* has to be necessarily interpreted as the subject of the *that*-clausesand of the infinitive clauses (in Italian). We can change this interpretation with N0 co-referent with the subject of the subordinate clauses by adding a different human noun (*Mary*):

(13) *Mike annuncia il fatto che Mary partirà a John*
      (*Mike announces the fact that Mary will be leaving to John*)
(14) *Mike comunica il fatto che Mary partirà a John*
      (*Mike communicates the fact that Mary will be leaving to John*)
(15) *Mike conferma il fatto che Mary partirà a John*
      (*Mike confirms the fact that Mary will be leaving to John*)

---

[4] A reduction for Harris is a transformation, which reduces the morphophonemic shape of a clause but leaves the lexical morphemes and the meaning unaltered.

3 The Paraphrases and the Dictionary

3.1 The Paraphrases

I have taken into account the following paraphrases:

- ProN0 = Pronominalization of N0
- ProN1 = Pronominalization of N1
- ProN2 = Pronominalization of N2
- RidN0 = Reduction of N0
- RidN2 = Reduction of N2
- PermN1 = Permutation of N1
- PermN2 = Permutation of N2
- Passivo = Passive sentence
- Vsup = Sentence with nominalization of the verb

These 9 type of paraphrases can be sequentially applied to the primitive sentence[5]. We can therefore produce combination of paraphrases like the following:

- S+PermN2+ProN2+RidN0
- S+PermN2+ProN2
- S+PermN2+RidN0
- S+PermN2
- S+RidN0+ProN2
- S+RidN0
- S+RidN0+RidN2
- S+RidN0+ProN2+Vsup1
- S+ RidN0+Vsup1
- S+RidN0+RidN2+Vsup1
- S+ProN2
- S+RidN2
- S+ProN2+Vsup1
- S+Vsup1

The previous combinations of paraphrases are a few among the dozens produced by the syntactic grammar. For example, the second paraphrase (S+PermN2+PronN2) in the list above can produce, when applied to a sentence like

I ragazzi comunicano il fatto che partono a Maria
(The boys communicate the fact that they are leaving to Maria)   (16)

a transformed sentence like the following:

---

[5] A primitive sentence is synonymous with lexicon-grammar definitional structure. It is different from the kernel, which is a sentence, unmarked in mood and voice (therefore is indicative and active).

$$A \text{ } lui \text{ } i \text{ } ragazzi \text{ } comunicano \text{ } il \text{ } fatto \text{ } che \text{ } partono$$
$$(To \text{ } him \text{ } the \text{ } boys \text{ } communicate \text{ } the \text{ } fact \text{ } that \text{ } they \text{ } are \text{ } leaving) \tag{17}$$

We see in sentence (17) the permutation of N2, a human noun, with its dislocation to the beginning of the sentence, and at the same time its pronominalization.

I have devised a syntactic grammar which can produce sentences either with a *that*-clause as N1 or with a noun in that same position. The noun according to the Harrisian grammar has to be interpreted as the reduction of a clause. The syntactic grammar has internal syntactic constraints which allow only certain types of paraphrases for each lexical entry in the dictionary associated with the syntactic grammar. The following is the syntactic grammar for the production of paraphrases for the Italian communication predicates:

The labels in bold type inside the graph refer to the types of paraphrases performed by the corresponding paths. Other types of paraphrases are hidden inside the embedded graphs (the yellow ones). The following two figures show the paraphrases automatically generated for the two types of sentences discussed above. Figure 2 represents the paraphrases with a noun in N1 position (direct object), while Fig. 3 represents a sentence with a clause in N1 position (*that*-clause):

Figures 2 and 3 above show examples of the different types of paraphrases produced by the syntactic grammar. The paraphrases, as already stressed in the introduction of this article, are sentences with a particular relationship with other sentences of which they have to be considered transformations. Harris points out very clearly what a transformation is in the following quotation:

"If two constructions (or sequences of constructions) occur with the same n-tuples of members of these classes in the same sentence environment (see below), we say that these constructions are transforms of each other, and that each may be derived from any other of them by a particular transformation. For example, the constructions N V v N (a sentence) and N's Ving N (a noun phrase) are satisfied by the same triple of N, V and N (*he, meet, we; foreman, put up, list,* and so on) ...: *He met us, his meeting us ...; The foreman put the list up, the foreman's putting the list up ...*".[6]

Besides being satisfied by the same n-tuples of members, two sentences which are transformations of each other have the same meaning. Each paraphrase automatically produced by NooJ and shown in Figs. 2 and 3 is followed by the name of the particular transformations applied in sequence. For example, the first paraphrase in Fig. 3 is followed by the label S+PermN2+ProN2+RidN0. S means that the syntactic grammar has recognized the paraphrase like a correct sentence; +PermN2 means that the complement N2 has been moved to a different position inside the sentence; +ProN2 means that the complement N2 has been pronominalized; +RidN0 refers to the dropping of the subject N0 (which is allowed in Italian but not in English).

The paraphrase just discussed is a transformation of what I have called a primitive sentence:

---

[6] Harris, Z.S.: Co-occurrence and transformation in linguistic structure. In: Language vol. 33, p. 288. Washington DC: Linguistic Society of America (1957).

I ragazzi annunciano il fatto chepartono a Maria
(The kids announce the fact that they are leaving to Maria) $\qquad$ (18)

A lei annunciano il fatto che partono
(To her they announce the fact that they are leaving) $\qquad$ (19)

In sentence (19) the subject *I ragazzi* (*The kids*) has been dropped (which is allowed in Italian); the N2 *a Maria* (*to Maria*) has been pronominalized to *a lei* (*to her*) and moved to the beginning of the sentence. The second paraphrase in Fig. 3 shows a different sequence of transformations:

A lei i ragazzi annunciano il fatto che partono
(To herthe kids announce the fact that they are leaving) $\qquad$ (20)

Compared with the primitive sentence (18), in sentence (20) one can notice the pronominalization of N2 *a lei* (*to her*) and its movement to the beginning of the sentence. Paraphrase (20) differs from paraphrase (19) because in the former no dropping of the subject has occurred. As I have already pointed out, the paraphrases now discussed preserve the meaning of the primitive sentence, in this case sentence (18).

Passive paraphrases have also been provided. In Fig. 3, a passive paraphrase occurs with no other transformation applied:

Il fatto che partono è annunciato dai ragazzi a Maria
(The fact that theyare leavingis announced by the kids to Maria) $\qquad$ (21)

Other passive sentences have been provided in association with other types of transformations. For example the passive transformation plus the pronominalization of N0 (+PronN0) and the dropping of N2 (+RidN2):

Il fatto che partono è annunciato da loro
(The fact that theyare leavingis announced by them) $\qquad$ (22)

Because Italian allows for the dropping of the subject, I have provided passive sentences without the N0. Various types of permutation of elements have been provided. For example, in Fig. 3 we find a paraphrase with the permutation of N2 (+PermN2):

A lei i ragazzi annunciano il fatto che partono
(To her the kids announce the fact that they are leaving) $\qquad$ (23)

Permutations of N1 can also occur in sequence with the pronominalization of N2 (+PermN1+ProN2):

Il fatto che partono i ragazzi annuncianoa lei
(The fact that theyare leavingthe kids announce to her) $\qquad$ (24)

As pointed out above, a different type of paraphrase allows for the occurrence of verbs in a nominalized form preceded by a carrier verb:

I ragazzi fanno un annuncio del fatto che partono a Maria
(The kids make the announcement of the fact that they are leaving to Maria).   (25)

The following embedded graph provides the occurrence of the carrier verbs in sentences like the (25):

The graph in Fig. 4 has six different "support verbs"[7], and each one of them is accepted by some communication predicates. In the dictionary related to this syntactic grammar I have added the properties "+FARE", "+DARE", and so on, which are syntactic constraints on the communication predicates of the dictionary

In order to produce paraphrases with the occurrence of nominalized verbs, I have devised the following embedded graph:

We see two types of determinants, *una* (*a*) for the nominalizations providing a feminine noun, and *un* (*a*) for the nominalizations providing a masculine noun.

3.2 The Dictionary

In order to produce correct paraphrases, the syntactic grammar shown in Fig. 1 needs to be associated with a dictionary. The dictionary is a list of all the communication predicates of lexicon-grammar class 47. Inside the syntactic grammar, an embedded graph has been provided whose main node contains the instruction V+47, allowing only the verbs listed in the associated dictionary to be parsed. The following figure shows an excerpt of the dictionary (Fig. 6):

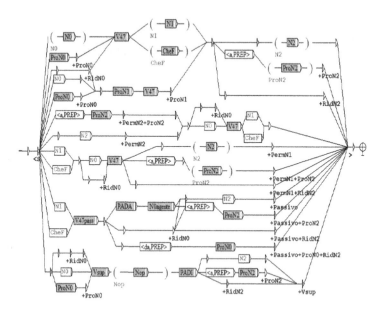

Fig. 1. The syntactic grammar for the Italian communication predicates (Color figure online)

---

[7] Maurice Gross has used the expression "support verb" which has to be considered synonymous with carrier verb.

```
a lei annunciano la partenza,S+PermN2+ProN2+RidN0
a lei i ragazzi annunciano la partenza,S+PermN2+ProN2
a Maria annunciano la partenza,S+PermN2+RidN0
a Maria i ragazzi annunciano la partenza,S+PermN2
annunciano la partenza a lei,S+RidN0+ProN2
annunciano la partenza a Maria,S+RidN0
annunciano la partenza,S+RidN0+RidN2
fanno un annuncio della partenza a lei,S+RidN0+ProN2+Vsup
fanno un annuncio della partenza a Maria,S+RidN0+Vsup
fanno un annuncio della partenza,S+RidN0+RidN2+Vsup
i ragazzi annunciano la partenza a lei,S+ProN2
i ragazzi annunciano la partenza a Maria,S
i ragazzi annunciano la partenza,S+RidN2
i ragazzi fanno un annuncio della partenza a lei,S+ProN2+Vsup
i ragazzi fanno un annuncio della partenza a Maria,S+Vsup
i ragazzi fanno un annuncio della partenza,S+RidN2+Vsup
i ragazzi la annunciano a lei,S+ProN1+ProN2
i ragazzi la annunciano a Maria,S+ProN1
i ragazzi la annunciano,S+ProN1+RidN2
la annunciano a lei,S+RidN0+ProN1+ProN2
la annunciano a Maria,S+RidN0+ProN1
la annunciano,S+RidN0+ProN1+RidN2
la partenza annunciano a lei,S+RidN0+PermN1+ProN2
la partenza annunciano a Maria,S+RidN0+PermN1
la partenza annunciano,S+RidN0+PermN1+RidN2
la partenza è annunciata a lei,S+RidN0+Passivo+ProN2
la partenza è annunciata a Maria,S+RidN0+Passivo
la partenza è annunciata da loro,S+Passivo+ProN0+RidN2
la partenza è annunciata dai ragazzi a lei,S+Passivo+ProN2
la partenza è annunciata dai ragazzi a Maria,S+Passivo
la partenza è annunciata,S+RidN0+Passivo+RidN2
la partenza i ragazzi annunciano a lei,S+PermN1+ProN2
la partenza i ragazzi annunciano a Maria,S+PermN1
la partenza i ragazzi annunciano,S+PermN1+RidN2
loro annunciano la partenza a lei,S+ProN0+ProN2
loro annunciano la partenza a Maria,S+ProN0
loro annunciano la partenza,S+ProN0+RidN2
```

Fig. 2. Paraphrases for sentences with a noun in N1 position

```
a lei annunciano il fatto che partono,S+PermN2+ProN2+RidN0
a lei i ragazzi annunciano il fatto che partono,S+PermN2+ProN2
a Maria annunciano il fatto che partono,S+PermN2+RidN0
a Maria i ragazzi annunciano il fatto che partono,S+PermN2
annunciano il fatto che partono a lei,S+RidN0+ProN2
annunciano il fatto che partono a Maria,S+RidN0
annunciano il fatto che partono,S+RidN0+RidN2
fanno un annuncio del fatto che partono a lei,S+RidN0+ProN2+Vsup
fanno un annuncio del fatto che partono a Maria,S+RidN0+Vsup
fanno un annuncio del fatto che partono,S+RidN0+RidN2+Vsup
i ragazzi annunciano il fatto che partono a lei,S+ProN2
i ragazzi annunciano il fatto che partono a Maria,S
i ragazzi annunciano il fatto che partono,S+RidN2
i ragazzi fanno un annuncio del fatto che partono a lei,S+ProN2+Vsup
i ragazzi fanno un annuncio del fatto che partono a Maria,S+Vsup
i ragazzi fanno un annuncio del fatto che partono,S+RidN2+Vsup
i ragazzi lo annunciano a lei,S+ProN1+ProN2
i ragazzi lo annunciano a Maria,S+ProN1
i ragazzi lo annunciano,S+ProN1+RidN2
il fatto che partono annunciano a lei,S+RidN0+PermN1+ProN2
il fatto che partono annunciano a Maria,S+RidN0+PermN1
il fatto che partono annunciano,S+RidN0+PermN1+RidN2
il fatto che partono è annunciato a lei,S+RidN0+Passivo+ProN2
il fatto che partono è annunciato a Maria,S+RidN0+Passivo
il fatto che partono è annunciato da loro,S+Passivo+ProN0+RidN2
il fatto che partono è annunciato dai ragazzi a lei,S+Passivo+ProN2
il fatto che partono è annunciato dai ragazzi a Maria,S+Passivo
il fatto che partono è annunciato,S+RidN0+Passivo+RidN2
il fatto che partono i ragazzi annunciano a lei,S+PermN1+ProN2
il fatto che partono i ragazzi annunciano a Maria,S+PermN1
il fatto che partono i ragazzi annunciano,S+PermN1+RidN2
lo annunciano a lei,S+RidN0+ProN1+ProN2
lo annunciano a Maria,S+RidN0+ProN1
lo annunciano,S+RidN0+ProN1+RidN2
loro annunciano il fatto che partono a lei,S+ProN0+ProN2
loro annunciano il fatto che partono a Maria,S+ProN0
loro annunciano il fatto che partono,S+ProN0+RidN2
loro fanno un annuncio del fatto che partono a lei,S+ProN0+ProN2+Vsup
```

Fig. 3. Paraphrases for sentences with a clause in N1 position

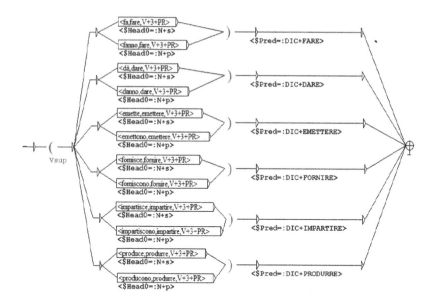

Fig. 4. Embedded graph for the paraphrases with carrier verbs

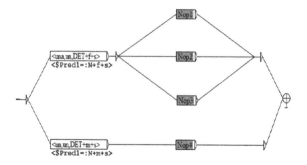

Fig. 5. Embedded graph for paraphrases with a nominalized verb

```
#classe 47
abbaiare,V+47+t+a+FLX=V11
abbozzare,V+47+t+a+FLX=V3
accennare,V+47+t+a+FLX=V3+DRV=ACCENNO:N5+Nop1+FARE
adombrare,V+47+t+a+FLX=V3
annunciare,V+47+t+a+FLX=V4+DRV=ACCENNO:N12+Nop2+FARE
ansimare,V+47+t+a+FLX=V3
articolare,V+47+t+a+FLX=V3
augurare,V+47+t+a+FLX=V3
balbettare,V+47+t+a+FLX=V3
barbugliare,V+47+t+a+FLX=V11
biascicare,V+47+t+a+FLX=V8
bisbigliare,V+47+t+a+FLX=V11
blaterare,V+47+t+a+FLX=V3
cablografare,V+47+t+a+FLX=V3+DRV=CABLOGRAFIA:N41+Nop3+FARE
cantare,V+47+t+a+FLX=V3+DRV=CANTATA:N41+Nop4+FARE
cantarellare,V+47+t+a+FLX=V3
canterellare,V+47+t+a+FLX=V3
cantilenare,V+47+t+a+FLX=V3+DRV=CANTILENA:N41+Nop5+FARE
certificare,V+47+t+a+FLX=V8+DRV=CERTIFICAZIONE:N46+Nop6+PRODURRE
chattare,V+47+t+a+FLX=V3
chiarificare,V+47+t+a+FLX=V3+DRV=CERTIFICAZIONE:N46+Nop7+DARE
chiarire,V+47+t+a+FLX=V201+DRV=CHIARIMENTO:N5+Nop8+DARE
chiedere,V+47+t+a+FLX=V33
citare,V+47+t+a+FLX=V3+DRV=CERTIFICAZIONE:N46+Nop9+FARE
citofonare,V+47+t+a+FLX=V3
```

Fig. 6. Excerpt from the dictionary

Besides the inflectional codes (+FLX=V11, +FLX=V3, and so on), derivational codes have been provided (+DRV=ACCENNO, +DRV=CANTATA, and so on). The derivational codes derive the nouns from the verbs allowing these nouns to inherit the syntactic property associated with the corresponding lexical entries. These codes for the derivation of nouns are followed by inflectional codes (N5, N12, and so on) whose role is that of producing the feminine and plural forms of the derived noun. The nouns derived from the verbs are useful in order to produce paraphrases with nominalized verbs.

At the end of each lexical entry of the previous dictionary, other syntactic properties have been added. These properties (+FARE, +PRODURRE, and so on) refer to the type of support verb allowed by the verb; they are present in the syntactic grammar as syntactic constraints because of the variation of syntactic form requested by the paraphrases according to the occurrence of different support verbs. As you can see in Fig. 5, only a small percentage of the communication predicates accept paraphrases with support verbs. Therefore, the syntactic grammar forbids the production of paraphrases with support verbs if the corresponding syntactic property is not found in the dictionary.

Some communication predicates can accept paraphrases with more than one support verb, with a statistic predominance of *fare* (*to make*) and *dare* (*to give*). The other support verbs are *emettere* (*to lodge*), *impartire* (*to teach*), *produrre* (*to show*), *fornire* (*to provide*) in support verb+communication predicate sequences like *emettere una contestazione* (*to lodge a complaint*), *impartire un insegnamento* (*to teach a lesson*), *produrre un document* (*to show a document*), *fornire un consiglio* (*to provide an advice*).

The following figure shows an excerpt of the dictionary containing some communication predicates which accept more than one support verb:

In Fig. 7, there are two verbs, *pronosticare* (*to prognosticate*) and *prospettare* (*to present*), which accept two different support verbs. *Pronosticare* (*to prognosticate*) accepts *fare* (*to make*) and *emettere* (*to lodge*), *prospettare* (*to present*) accepts *fare* (*to make*) and *dare* (*to give*). You find the support verbs as already shown above, added to the verbs as syntactic properties (+FARE+EMETTERE, +FARE+DARE).

```
notificare,V+47+t+a+FLX=V8+DRV=CANTILENA:N41+Nop42+DARE
obiettare,V+47+t+a+FLX=V3
ordinare,V+47+t+a+FLX=V3+DRV=MENZIONE:N5+Ncp43+DARE+CONG
postulare,V+47+t+a+FLX=V3+DRV=CERTIFICATO:N5+Ncp44+FARE
preaccennare,V+47+t+a+FLX=V3
preannunciare,V+47+t+a+FLX=V4+DRV=ACCENNO:N12+Nop45+FARE
precisare,V+47+t+a+FLX=V3+DRV=CERTIFICAZIONE:N46+Nop46+FARE
preconizzare,V+47+t+a+FLX=V3+DRV=CERTIFICAZIONE:N46+Nop47+FARE+FUT
predicare,V+47+t+a+FLX=V8+DRV=CERTIFICAZIONE:N46+Nop48+FARE
predire,V+47+t+a+FLX=V217+DRV=CERTIFICAZIONE:N46+Nop49+FARE
premettere,V+47+t+a+FLX=V28+DRV=PREMESSA:N41+Nop50+FARE
professare,V+47+t+a+FLX=V3+DRV=CONFESSIONE:N46+Ncp51+FARE
promettere,V+47+t+a+FLX=V28+DRV=PREMESSA:N41+Nop52+FARE
pronosticare,V+47+t+a+FLX=V8+DRV=ACCENNO:N5+Nop53+FARE+EMETTERE
proporre,V+47+t+a+FLX=V240
prospettare,V+47+t+a+FLX=V3+DRV=ACCENNO:N5+Nop54+FARE+DARE
raccomandare,V+47+t+a+FLX=V3+DRV=CERTIFICAZIONE:N46+Nop55+CONG
raccontare,V+47+t+a+FLX=V3+DRV=ACCENNO:N5+raccontoNop56+FARE
radiotrasmettere,V+47+t+a+FLX=V28+DRV=RADIOTRASMISSIONE:N46+Nop57+DARE
raffigurare,V+47+t+a+FLX=V3+DRV=CERTIFICAZIONE:N46+Nop58+FARE
rammentare,V+47+t+a+FLX=V3
rapportare,V+47+t+a+FLX=V3+DRV=ACCENNO:N5+Nop59+FARE
ratificare,V+47+t+a+FLX=V8+DRV=CANTILENA:N43+Nop60+DARE
recitare,V+47+t+a+FLX=V3+DRV=CANTILENA:N41+Nop61+FARE
relazionare,V+47+t+a+FLX=V3+MENZIONE:N46+Nop62+FARE
rettificare,V+47+t+a+FLX=V8+DRV=CANTILENA:N43+Nop63+FARE
riassumere,V+47+t+a+FLX=V58+DRV=RIASSUNTO:N5+Nop64+FARE
```

Fig. 7. Excerpt from the dictionary

4   Conclusions

More work needs to be done in order to take into account the production of paraphrases with Vsup+nouns. The set of support verbs accepted by the communication predicates of class 47 can be increased. When new support verbs are added to the dictionary, the corresponding syntactic grammar needs to undergo some changes by adding new paths and discriminating between the previous ones. Researches have showed that sentences with nominalizations could be half the total number of the sentences belonging to a lexicon-grammar class.

Each communication predicate is syntactically different from the others, which means that the syntactic grammar devised here needs to be improved by implementing specific syntactic constraints. The constraints can discriminate between each communication predicate in order to produce ad-hoc paraphrases according to the lexicon-grammar differences between the communication predicates of class 47. I am also working on the possibility of increasing the number of paraphrases produced by the syntactic grammar.

References

1. Elia, A., Langella, A.M.: Semantic role labelling with NooJ: communication predicate in Italian. In: Formalizing Natural Languages with NooJ 2014, pp. 76–86. Cambridge Scholars Publishing, Newcastle upon Tyne (2015)
2. Elia, A., Martinelli, M., D'Agostino, E.: Lessico e Strutture della Sintassi. Liguori Editore, Napoli (1981)
3. Gross, M.: Méthodes en Syntaxe. Hermann, Paris (1975)
4. Gross, M.: Une Grammaire Locale de l'Expression des Sentiments. Langue Française **105**, 70–87 (1995). Larousse, Paris
5. Harris, Z.S.: Mathematical Structures of Language. Interscience Publishers, New York (1968)
6. Harris, Z.S.: Language and Information. Columbia University Press, New York (1988)
7. Harris, Z.S.: A Grammar of English on Mathematical Principles. Interscience Publishers, New York (1982)
8. Harris, Z.S.: Papers in Structural and Transformational Linguistic. D. Reidel Publishing Company, Dordrecht (1970)
9. Kawahara, D., Shinzato, K., Shibata, T., Kurohashi, S.: Precise information retrieval exploiting predicate-argument structures. In: International Joint Conference on Natural Language, Asian Federation of Natural Language Processing, pp. 37–45, Nagoya (2013)
10. Koeva, S., Maurel, D., Silberztein, M.: Intex et NooJ pour la Linguistique et le Traitement Automatique des Langues. In: Formalizer les Langues avec l'Ordinateur: de Intex à NooJ, pp. 9–16. Presses Universitaires de Franche-Compté, Besançon (2007)
11. Langella, A.M.: Elementi di grammatica dell'italiano su basi matematiche. Libreriauniversitaria.it, Padova (2014)
12. Langella, A.M., Messina, S.: Paraphrases V↔N↔A in one class of psychological predicates. In: Formalizing Natural Languages with NooJ 2014, pp. 140–151. Cambridge Scholars Publishing, Newcastle upon Tyne (2015)
13. Lentin, A.: Reflections on references to mathematics in the work of Zellig Harris. In: The Legacy of Zellig Harris. John Benjamins Publishing Company, Philadelphia. (2002)
14. Silberztein, M.: NooJ's Linguistic Annotation Engine. In: Formalizer les Langues avec l'Ordinateur: de Intex à NooJ, pp. 17–34. Presses Universitaires de Franche-Compté, Besançon (2007)
15. Silberztein, M.: Syntactic parsing with NooJ. In: Proceedings of the NooJ 2009 International Conference and Workshop, pp. 177–190. Centre de Publication Universitaire, Tunis (2010)
16. Silberztein, M.: Automatic Transformational Analysis and Generation. In: Proceedings of the 2010 International NooJ Conference, pp. 221–231. Komotini: Democritius University Editions, Thrace (2011)
17. Silberztein, M.: Analyse et Génération Transformationnelle avec NooJ. In: Proceedings of the 47th Annual Meeting of the Italian Linguistic Society, pp. 234–252. Bulzoni Editore, Roma (2016)
18. Silberztein, M.: Formalizing Natural Language: the NooJ Approach. Wiley, London (2016)
19. Vietri, S.: Transformations and Frozen Sentences. In: Proceedings of 2011 NooJ International Conference and Workshop, pp. 166–180. Cambridge Scholars Publishing, Newcastle upon Tyne (2012)

eSPERTo's Paraphrastic Knowledge Applied to Question-Answering and Summarization

Cristina Mota[⊠], Anabela Barreiro, Francisco Raposo,
Ricardo Ribeiro, Sérgio Curto, and Luísa Coheur

INESC-ID, Rua Alves Redol 9, 1000-029 Lisbon, Portugal
{cmota,francisco.afonso.raposo}@ist.utl.pt,
{anabela.barreiro,ricardo.ribeiro,sergio.curto,
luisa.coheur}@inesc-id.pt

Abstract. This paper reports our first attempt of integrating eSPERTo's paraphrastic engine, which is based on NooJ platform, with two application scenarios: a conversational agent, and a summarization system. We briefly describe eSPERTo's base resources, and the necessary modifications to these resources that enabled the production of paraphrases required to feed both systems. Although the improvement observed in both scenarios is not significant, we present a detailed error analysis to further improve the achieved results in future experiments.

Keywords: Paraphrasing · Question-Answering · Summarization · Portuguese · NooJ

1 Introduction

eSPERTo is a paraphrasing system that comprises a paraphrase generator, a paraphrase acquisition module, and a web interactive application to help Portuguese language learners, translators and editors in revising their texts. This system was developed in the scope of the eSPERTo project whose aim is to build a hybrid paraphrasing system. Its core linguistic resources, extracted from OpenLogos bilingual resources [3], the free open source version of the Logos System [15], were adapted and integrated into NooJ linguistic engine [16].

In this paper, we present a study on the integration of the paraphrase generator into two application scenarios: a conversational agent, and a summarization system. In the first application scenario, eSPERTo's paraphrases were explored to enrich the Portuguese knowledge base of a question-answering intelligent virtual conversational agent, EDGAR [5]. In the second application scenario, eSPERTo was used in the pre-processing phase to assist automatic text summarization [12].

The benefits of paraphrastic knowledge to these natural language processing tasks have been defined and quantified previously. One the one hand, discovering paraphrased answers provides additional evidence that an answer is correct and helps systems to identify semantically related questions for the same answer [14], and, on the other hand, the identification of paraphrases allows information across documents to be

© Springer International Publishing AG 2016
L. Barone et al. (Eds.): NooJ 2016, CCIS 667, pp. 208–220, 2016.
DOI: 10.1007/978-3-319-55002-2_18

condensed, redundant information to be identified and eliminated and helps improve the quality of the generated summaries [4].

2 The eSPERTo's Paraphrasing System

The eSPERTo's paraphrasing system consists of three main components: a paraphrase generator, a paraphrase acquisition module, and a web interface. Figure 1 depicts the general architecture of eSPERTo. The paraphrase generator is based on NooJ technology. More specifically, the resources conform to NooJ format, have been mostly developed with the NooJ window environment, and then are applied to texts with NooJ linguistic engine.

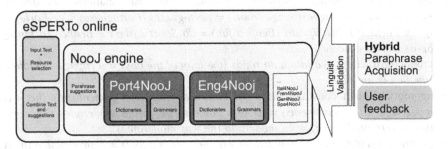

Fig. 1. Architecture of eSPERTo's paraphrasing system

2.1 eSPERTo's Resources

eSPERTo uses the Portuguese resource module of NooJ [2], extracted and adapted from the OpenLogos bilingual resources [3]. This module already included resources that help generate paraphrases involving, among others:

- support verb constructions or their stylistic or aspectual variants and corresponding single verbs – *fazer/realizar/efetuar uma apresentação* (*make a presentation* (*of*)) = *apresentar* (*to present*)
- compound and single adverbs – *de uma forma interativa* (*in an interactive way*) = *interativamente* (*interactively*); *com entusiasmo* (*with enthusiasm*) = *entusiasticamente* (*enthusiastically*)
- relatives and participial adjectives – *que foram escritos* (*that were written*) = *escritos* (*written*)
- relatives and possessive constructions – *o papel que a Europa tem/desempenha* (*the role that Europe plays*) = *o papel da Europa* (*the role of Europe*)
- active/passive constructions – *A solta B* (*A releases B*) = *B é solto por A* (*B is released by A*)

The most recent version of Port4Nooj [8] includes three new components: (i) a lexicon-grammar based dictionary of human intransitive adjectives, and a set of local grammars that use the distributional properties of those adjectives for paraphrasing,

(ii) a polarity dictionary for sentiment analysis, and (iii) a set of priority dictionaries and local grammars for named entity recognition.

However, at the time of the study presented in this paper, the Port4NooJ module had not been updated yet with neither the polarity lexicon nor the resources for named entity recognition, it had only been updated with the lexicon-grammar based dictionary of human intransitive adjectives [9]. In the latter study, the integration of the lexicon-grammar tables describing the syntactic and distributional properties of these adjectives allows, among others, the paraphrasing of:

- adjective, noun and verb morphologically related constructions – *está alegre* (*is happy*) = *alegrou-se* (*got* (*self*) *happy*) = *sentiu alegria* (*felt happiness*)
- adjective constructions supported by different copulative verbs – *estar perdido* (*to be lost*) = *andar perdido* (*walk around lost*)
- constructions involving nationality and other membership relations – *de origem portuguesa* (*of Portuguese origin/roots*) = *portugueses* (*Portuguese*) = *de Portugal* (*from Portugal*); *benfiquista* (*Benfica fan*) = *do Sport Lisboa e Benfica* (*a fan of Sport Lisboa e Benfica*)
- cross-constructions – *o idiota do rapaz* (*the idiot of the boy*) = *o rapaz é um idiota* (*the boy is an idiot*)
- appropriate noun constructions – *foi moderado nos seus comentários* (*he was moderated in his comments*) = *os seus comentários foram moderados* (*his comments were moderated*) = *foi moderado* (*he was moderated*)
- generic noun phrases – *é um indivíduo estúpido* (*he is a fool*) = *é um estúpido* (*he is a fool*) = *é estúpido* (*he is a fool*)
- characterizing indefinite constructions, where an indefinite article precedes the adjective – *o rapaz é um idiota* (*the boy is an idiot*) = *o rapaz é idiota* (*the boy is idiot*).

2.2   Paraphrase Processing

eSPERTo's web interface, represented in Fig. 2, translates the parameters chosen by the user in a request to NooJ, which is parameterized with the appropriate lexical and syntactic resources as seen in Table 1, where *Language* represents the input language, *Output* represents the output file with the initial and ending positions of the sequences to be rewritten followed by the rewritten sequences, *LexicalR* represents one or more lexical resources (either dictionaries or morphological grammars), *SyntacticR* represents one or more syntactic grammars that may add increasing levels of annotations, *Annotations* is the final grammar, which will produce the rewritten sequences, and *Input* represents the input text.

The first set of syntactic grammars, *SyntacticR*, that adds increasing levels of annotations, is applied sequentially in *longest mode*. Each grammar may add annotations that are used by the grammars applied subsequently to recognize more complex sequences. Some of these grammars identify sequences that can be paraphrased, and they also delimit/associate these sequences with one or more annotations with the label REESCREVE, that means, *rewrite*. One of the attributes of this label is TEXTO, whose

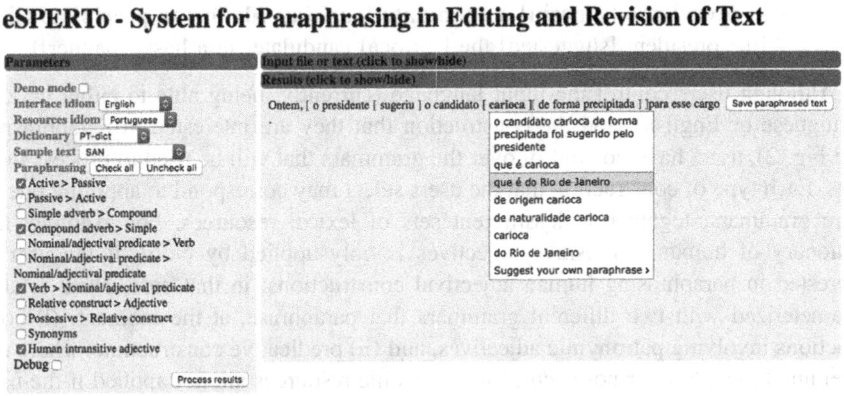

eSPERTo - System for Paraphrasing in Editing and Revision of Text

Fig. 2. eSPERTo's web interface

Table 1. NooJ arguments for different user paraphrasing requests

	Active → Passive	Verb → N/Adj Predicate
Language	pt	pt
Output	file1.ind	file2.ind
LexicalR	PT-dict.nod	PT-Dict.nod
	PT-Contr.nom	PT-Contr.nom
SyntacticR	GN.nog	GN.nog
	g00_PARA_ACT2PASS.nog	g04_PARA_APV2 V.nog
		g04_PARA_PAC2 V.nog
		g04_PARA_SVC2 V.nog
		g04_PARA_VSTYLE2 V.nog
Annotations	_REESCREVE-A.nog	_REESCREVE-A.nog
Input	file1.txt	file2.txt

value is one of the paraphrases of the matched sequence. The sequence will be delimited by as many annotations as the number of equivalent paraphrases that the grammar produces.

The last grammar, *Annotations*, is applied in all sequences mode and outputs the rewritten sequences by consulting the value of the attribute TEXTO of all annotations REESCREVE. This grammar is not applied in longest mode, because the longest sequence annotated may include smaller subsequences that can also be paraphrased. In this way, as illustrated in Fig. 2, eSPERTo's paraphrase generator suggests paraphrasing the longest match shown in example (1), which is an active construction to its passive equivalent, but also (i) the verb *sugeriu* (*suggested*) to the equivalent predicative noun construction *fez a sugestão* (*made the suggestion*), (ii) the adjective *carioca* to, for example, *do Rio de Janeiro* (*from Rio de Janeiro*), and (iii) the compound adverb *de forma precipitada* (*in a hasty manner*) to the simple adverb *precipitadamente* (hastly).

(1) $_{PT}$ – [**o presidente [sugeriu] o candidato [carioca] [de forma precipitada]**]
   $_{EN}$ – [the president [suggested] the [carioca] candidate [in a hasty manner]]

Although users control the input language (currently, being able to either choose Portuguese or English), and the construction that they are interested in paraphrasing (see Fig. 2), users have no control over the grammars that will be used to process their texts. Each type of construction that the users select may correspond to applying one or more grammars, together with different sets of lexical resources. For example, the dictionary of human intransitive adjectives is only applied by NooJ, if the user is interested in paraphrasing human adjectival constructions; in that case, NooJ is also parameterized with two different grammars that paraphrase, at the moment, (i) constructions involving patronymic adjectives, and (ii) predicative constructions where the indefinite article may or not occur. Those specific resources are not applied if the user only opts for one or more of the remaining possibilities. On the other hand, the grammar to identify noun phrases is applied regardless of the selection made by the user. Table 1 illustrates two other examples of NooJ parameterization. As can be seen, in the first case only one paraphrasing grammar is applied, but in the second case, the request to paraphrase verbs into nominal and adjectival constructions results in applying four different grammars (grammar names started by g04).

3   eSPERTo in Question-Answering

In the first application scenario, the paraphrases generated by eSPERTo were explored to enrich the knowledge base of an intelligent virtual conversational agent, EDGAR. EDGAR answers questions asked by the visitors of a Portuguese museum, the Palácio de Monserrate in Sintra. Its knowledge base is built on question/answer pairs. Given a user utterance, EDGAR calculates its lexical distance to each question in the knowledge base; the question with the shortest distance to the user utterance will trigger the respective answer.

In compile-time, EDGAR's knowledge base is extended with a list of hand-crafted equivalence relations. The paraphrase generator allows to automatically enrich this list by providing all possible ways of rewriting the same question, thereby enabling the same answer to semantically equivalent questions. For the question *És português?* (*Are you Portuguese?*), eSPERTo takes the word *português* and generates the equivalent phrases: *de Portugal* (*from Portugal*) and *de nacionalidade portuguesa* (*of Portuguese nationality*). Thus, the following alternative equivalent questions were added to EDGAR's knowledge base:

És de Portugal?
És de origem portuguesa?
És de nacionalidade portuguesa?

3.1   Results

EDGAR's knowledge base had originally 848 sentences. eSPERTo matched 2,028 times with sequences from these sentences, being 359 unique matches. To avoid looping during the expansion of the knowledge base, some paraphrases, such as *ingleses/que são ingleses (English/that are English)*, were discarded. The resulting list was added to EDGAR's equivalence relation list, without further processing. Then, results were compared with the baseline. Recall improved (from 0.7972 to 0.8149) by hurting precision, and although F-measure increased, improvements are not significant (see Table 2).

Table 2. EDGAR's performance with and without eSPERTo

	Recall	Precision	F-measure
Baseline	0.7972	0.7889	0.7930
Baseline + eSPERTo	0.8149	0.7763	0.7951

3.2   Error Analysis

The insignificant improvement was not surprising, especially due to two reasons. First, eSPERTo was not developed with the specific purpose of generating paraphrases within the context of questions, which introduces a different level of analysis that the grammars need to be able to process. Second, many grammars are not yet generating the correct inflected form.

Incorrect Inflection of Paraphrases. Paraphrases do not always keep the same inflection of their triggers. The grammars that paraphrase verbal constructions into their predicative adjective or noun equivalent constructions (and vice-versa) generate the infinitive form of the auxiliary or support verb (or the main verb, when rephrasing from the construction with support verb to the verbal construction) instead of generating the corresponding form that triggered the rephrasing. This was an initial approach, as, in NooJ syntactic grammars, generating the inflection of an output word according to another input word is not yet straightforward, and we wanted a simple way of validating this type of paraphrasing. We are currently working in generating the correct forms of all paraphrased verbs.

Suggesting to the user of the eSPERTo's web interface a paraphrase in the infinitive form is not a serious problem, because the user can always accept the suggested paraphrase and then correct the inflection. However, for an automatic system, such as EDGAR, syntactically incorrect paraphrased sentences will produce no match in any input utterance. For example, for the question *Vives onde? (Where do you live?)*, eSPERTo found the suitable support verb construction *fazer vida (to make life)* for *viver (to live)*, but did not inflect the support verb *fazer* with the same inflectional features of *vives ((you) live)*, which is in the second person singular of the present tense. This resulted in an unwanted, and unintentional integration of the syntactically incorrect question *Fazer vida onde? (to make life where?)* into EDGAR's database. Section 3.2 shows that errors of this kind do not occur in the summarization case scenario. The reason for this is that, for now, we used a subset of all the grammars applied in EDGAR, which did not include verbal constructions that can be equivalent to support verb constructions.

Incorrect Rephrasing due to Ambiguity. Another challenge that results in errors is the ambiguity between words. Some errors that occur in eSPERTo's rephrasing sequences would not show up if the words had been disambiguated in advance. One such example is *como* in Portuguese. This word can either be the interrogative pronoun *how/what*, the conjunction or preposition *as* or a form of the verb *to eat*. In the context of EDGAR, it is never the verb, but, for the question *Como te chamas? (What is your name?)*, eSPERTo incorrectly suggests the support verb construction *tomar comida (to have food)* for the word *como*, recognizing it as a verb instead of an interrogative pronoun. An easy solution to overcome this problem is to add the entries for *como* as a pronoun, preposition and conjunction to a priority dictionary, preventing the main dictionary of analyzing it also as verb. Consequently, the grammar that generates the support verb constructions will not match occurrences of *como* in EDGAR's database.

In more complicated cases, where the verb is not syntactically ambiguous, eSPERTo needs disambiguation grammars or use STRING's disambiguation component before applying its rephrasing grammars.

Incorrect Rephrasing due to Missing Multiwords. For example, for the question *Qual é que é o seu nome completo? (What is your full name?)*, eSPERTo suggests [*completo = ficar completo*] ([*complete = to remain/become/be complete*]), because it does not recognize the multiword unit *nome completo (full name)*. Once the multiword unit is lexicalized, the ambiguity disappears and the verb *ficar (to remain, to become, to be)* will be eliminated, because the adjective *completo* will no further be identified as a verbal (participial) adjective related to the verb *completar*, to which the support verb construction *ficar completo*, and subsequently, the adjective *completo* are semantico-syntactically related.

4   eSPERTo in Summarization

In the second application scenario, eSPERTo was used to assist an automatic text summarization module. This module [12, 13] was developed in the context of SSNT (*Sumarização de Serviços Noticiosos Televisivos*), a prototype developed for the selective dissemination of multimedia content [1, 10, 17]. More specifically, eSPER-To's paraphrases were used in the pre-processing phase of the summarization component [12] to detect redundant content in TeMário, a corpus of 100 newspaper articles in Brazilian Portuguese [11].

This means that instead of providing the summarization system with all the possibilities of rewriting the same sequence, as in the case of the conversational agent, eSPERTo rewrites different phrases that are equivalent with the same paraphrase. In this way, if we have, for example *o presidente dos Estados Unidos discursou nas Nações Unidas (the president of the United States spoke at the United Nations)* and also *o dirigente americano viajou para o Iraque (the American leader traveled to Iraq)* in the same text, when eSPERTo rephrases *dos Estados Unidos* with *americano*, the summarizer should more easily identify the co-reference between *American president* and *American leader*, hence, helping in generating a shorter version of the text, for example, *o presidente americano discursou nas Nações Unidas e viajou para o Iraque (the American president spoke at the United Nations and traveled to Iraq)*.

The main challenge is to identify the best candidate among the equivalent expressions to be used in rewriting the text. We evaluated four different groups of paraphrases: (i) active/passive, (ii) constructions involving patronymic adjectives (is it better to use the shortest construction *o prefeito carioca* (*the carioca Mayor*) or the one with the equivalent toponym, *o prefeito do Rio de Janeiro* (*the Mayor of Rio de Janeiro*)?), (iii) simple adverb (*rapidamente – quickly/fast*) equivalent adjectival construction (*de forma rápida – in a quick/fast way*), (iv) adjectival (*o presidente ficou desiludido – the president got disappointed*), nominal (*a desilusão do presidente – the disappointment of the president*) and reflexive verbal predicates (*o presidente desiludiu-se – the president got* (*himself*) *disappointed*). This corresponds to generating different versions of TeMário by applying different paraphrasing grammars.

4.1 Results

Table 3 shows preliminary statistics of applying a subset of eSPERTo's grammars to pre-process texts in TeMário. They were applied after normalization. As illustrated in the table, 90% of the documents have at least one sequence that was rephrased when the grammars were applied to rephrase (i) nationality adjectives, (ii) compound adverbs into adverbs ending with suffix *-mente* (*-ly*), and (iii) passive into active. However, some of those grammars either (i) produce an undefined result, because when they try to create the output some features are missing from the dictionary entries, or (ii) produce the same sequence as the sequence already in the text, as it happens with patronymic adjectives when the construction matched in the text is the one chosen to represent all paraphrases of that construction. If the sequences that only produce undefined cases are not counted, the number of documents with sequences that are rephrased decreases to 74, and if additionally the sequences that are replaced by the same sequence are not counted in, the number of documents decreases to 62.

Table 3. Rewriting statistics before summarization

ID	Paraphrasing	# Docs	# Sequences
1	ADVmente → (de modo|maneira|jeito A)|(com N)	80	215
2	(de modo|maneira|jeito A) → ADVmente	73 (26/26)	322
3	SAN	70 (70/48)	305
4	Passive → Active	7	7
5	Active → Passive	33	58
2, 3, 4	Shortest paraphrases	90 (74/62)	682

ROUGE-1 [6] was used to evaluate the new summaries generated after rephrasing, and the scores were compared to the ones without using eSPERTo reported in the literature [13].

Table 4 shows the scores for the paraphrases that improved most significantly.

Table 4. Summarization evaluation

System	Rephrasing ID	ROUGE-1
Manhattan (SSC = 2)	5	0.444
Fractional (N = 1.(3), SSC = 2)	5	0.443
Fractional (N = 1.(3), SSC = 2)	1	0.443
Fractional (N = 1.(3), idf, H1.3)	4	0.443
Fractional (N = 1.(3), SSC = 2)	–	0.442
Fractional (N = 1.(3), idf, H1.3)	–	0.442
Manhattan (SSC = 2)	–	0.442

4.2   Error Analysis

Table 5 shows text St-mu94ab03-a of TeMário before and after being rephrased by eSPERTo prior to summarization. The rephrased sequences serve to illustrate some of the problems encountered and possible solutions to fix those problems.

Table 5. Text St-mu94ab03-a before and after rephrasing aiming at text shortening

Quais seriam as reações desejáveis no campo macroeconômico por parte das autoridades **da Europa e do Japão**. (…) Com uma renda maior os norte-americanos oferecerão mercados de exportação mais fortes para a Europa em geral para os países do Sudeste **Asiático** e para o mundo em vias de desenvolvimento. (…) **Com frequência** cada vez maior as ações e os bônus parecem mover-se juntos. Quando os bancos centrais derrubam os preços dos títulos as ações tendem a acompanhá-los mesmo quando **o aperto do crédito foi desencadeado pela perspectiva** de lucros e produção em alta. (…) E se vários países atuassem em conjunto Espanha Itália França e Reino Unido uma modesta apreciação do dólar melhoraria a competitividade européia **de maneira muito oportuna**.

Quais seriam as reações desejáveis no campo macroeconômico por parte das autoridades **europeias e nipónico**. (…) Com uma renda maior os norte-americanos oferecerão mercados de exportação mais fortes para a Europa em geral para os países do Sudeste **asiático** e para o mundo em vias de desenvolvimento. (…) **Frequentemente** cada vez maior as ações e os bônus parecem mover-se juntos. Quando os bancos centrais derrubam os preços dos títulos as ações tendem a acompanhá-los mesmo quando **a perspectiva de lucros desencadeou o aperto do crédito** e produção em alta. (…) E se vários países atuassem em conjunto Espanha Itália França e Reino Unido uma modesta apreciação do dólar melhoraria a competitividade européia **oportunamente**.

Lack of Agreement between Noun and Adjective. In the noun phrase *as autoridades da Europa e do Japão* (*the authorities from Europe and Japan*), although eSPERTo generates the correct feminine plural form of the patronymic adjective corresponding to *da Europa* (*from Europe*), *europeias* (*European*), it failed to do the same with *nipónicas*, generating instead the masculine singular form when rephrasing *do Japão* (*from Japan*) with the adjective *nipónico* (*Japanese*). This happens because, so far, the grammar is not describing coordination, and therefore agreement between all the elements of the coordination and the head noun is not achieved yet. On the other hand, if the head noun of the noun phrase is plural, one may want to generate the singular forms for each form involved in the coordination, such as, in the noun phrase *Os presidentes da China Jiang Zemin e da Rússia Boris Yeltsin* (*the presidents of China Jiang Zemin and of Russia Boris Yeltsin*) that should generate the paraphrase *Os presidentes chinês Jiang Zemin e russo Boris Yeltsin* (*the Chinese president Jiang Zemin and Russian Boris Yeltsin*).

Other reasons for the lack of agreement are: (i) the head noun may be ambiguous between a singular and plural form, or between a masculine and a feminine form, or (ii) the patronymic adjective may be incorrectly associated with another adjective preceding it, which is ambiguous with a noun, instead of being associated with the head of the noun phrase. An example of the latter case occurs with *os bancos centrais da Europa* (*the central banks of Europe*), where *da Europa* is rephrased with *europeias* (the feminine form of *European*) because *centrais* besides being the plural form of a neutral adjective *central*, is also the feminine plural noun *plants*, as in *centrais nucleares* (*nuclear plants*). This problem, depending on the nature of the sequence involved, may be solved in different ways. In this particular case, *banco central* is a multiword noun, and, if identified as such by a multiword lexicon, the adjective *europeu* would be unambiguously modifying this noun, hence, it would only generate the correct masculine form for this phrase. Another way, which in any case should refine rephrasing, is to include a disambiguation phase prior to rephrasing in order to process *centrais* as masculine only.

Invalid Rephrasing within Named Entities. Rephrasing cannot occur within named entities, or it is only allowed in specific cases. For example, in the named entity *Sudeste Asiático* (*Southeast Asia*), the patronymic adjective *Asiático* (*Asian*) can be suitably rephrased with *da Ásia* (*of Asia*). However, in similar cases the rephrasing should not take place. For example, in the named entity *Banco do Brasil* (*Bank of Brazil*), *do Brasil* (*of Brazil*) should not be rephrased with *brasileiro* (*Brazilian*), as it alters the name of the entity. Nonetheless, rephrasing could be explored from the named entity to a common noun phrase, for example, to *banco brasileiro* (*Brazilian bank*). This rephrasing, however, introduces ambiguity, as it can refer to any other Brazilian bank, in addition to the *Banco do Brasil*, a specific organization.

Incorrect Delimitation of the Sequence to Rephrase. In some cases, the grammars are not fully delimiting the sequence to rephrase, resulting in incorrect rephrasing. For example, *com frequência* (*with frequency*), in Table 5, was identified as the sequence to rephrase, but *com frequência cada vez maior* (*with increasingly more frequency*) is the correct sequence to be rephrased into *cada vez mais frequentemente* (*increasingly more frequently*), a paraphrase more complex than the ones currently achieved by eSPERTo's grammars.

Rephrasing from active to passive or vice-versa also requires improvement in terms of delimiting the subject and object of the construction. In sentence (2), the subject of *desencadeado* (*triggered*) is the noun phrase *a perspectiva de lucros e produção em alta* (*the prospect of profits and high production*), but the passive construction is incorrectly rephrased as *a perspectiva desencadeou o aperto do crédito de lucros e produção em alta* (*the prospect triggered the credit squeeze of profits and high production*), as illustrated in the incorrect paraphrase (*PARA*) represented in example (2).

(2) *PT* –... as ações tendem a acompanhá-los mesmo quando **o aperto do crédito foi desencadeado pela perspectiva de lucros e produção em alta**

    EN – the actions tend to accompany them even when the credit squeeze was triggered by the prospect of profits and high production

    PARA –... *as ações tendem a acompanhá-los mesmo quando **a perspectiva desencadeou o aperto do crédito** de lucros e produção em alta.

For an easier and adequate parsing of the noun phrases that can be subject or object of the active-passive constructions, eSPERTo also needs to improve the detection of multiwords, as in example (3), and of named entities, as in (4).

(3) *PT* – **A decisão foi anunciada pelo secretário** de estado americano Warren Christopher pouco antes dele iniciar sua primeira viagem oficial à América Latina.

    EN – The decision was announced by U.S. Secretary of State Warren Christopher just before he started his first official trip to Latin America.

    PARA – ***o secretário anunciou A decisão** de estado americano Warren Christopher pouco antes dele iniciar sua primeira viagem oficial à América Latina.

In example (3), *secretário de estado* (*Secretary of State*) was not recognized as a compound noun followed by the name of the Secretary of State.

(4) *PT* – que no **seu caso foi invocada pelo Movimento Contra** o Racismo e pela Amizade entre os Povos.

    EN – which in his case was invoked by the Movement Against Racism and for the Friendship between Peoples.

    PARA – que no **o Movimento Contra invocou seu caso** o Racismo e pela Amizade entre os Povos.

In example (4), the named entity *Movimento Contra o Racismo e pela Amizade entre os Povos* (*Movement Against Racism and for the Friendship between Peoples*) was not identified as such, both resulting in incorrect rephrasing evidenced by the application of a passive-to-active grammar.

Incorrect Identification of the Sequence to Rephrase. eSPERTo's grammars also need to be more precise in the contexts where the rephrasing can take place. For example, the rephrasing of *com influência* (*with influence*) with *influentemente* (*influentially*) should not occur when the adjective is modified by another adjective, as in the phrase *Com influência inicial do expressionismo alemão* (*With initial influence of the German expressionism*).

Another common case, where paraphrasing should not have occurred is when the verb does not allow the passive construction, as the verb *confundir* in example (5).

(5) *PT* – Começou por considerar **uma calúnia confundir teu livro com as teses revisionistas**.

    EN – (He/she) started by considering a defamation to confuse your book with the revisionist theses.

PARA – Começou por considerar **teu livro com as teses revisionistas ser con-fundido por uma calúnia.**

Example 5 reinforces the importance of including an attribute to the verbal entries indicating whether the verb accepts the passive construction. Ideally, one should have a lexicon-grammar for verbs, which would include this and other relevant information.

5   Conclusions and Future Work

After integrating the eSPERTo's paraphrase generator with a conversational agent and a summarization system, minor improvements in performance in both scenarios were observed. This integration was done before a thorough evaluation of eSPERTo's paraphrasing capabilities. However, by analyzing the paraphrases obtained in these two scenarios, we did an extrinsic evaluation of eSPERTo pinpointing different types of problems and solutions to further improve eSPERTo's paraphrasing "smartness". Some of the problems observed are specific to the application scenario and, hence, will require domain specific solutions.

As the preliminary results show, paraphrase generation could be refined by using a disambiguation component and a named entity tagger. We plan to integrate eSPERTo with STRING [7], an NLP system developed at INESC-ID, which already performs disambiguation, chunking parsing, named entity recognition, among other tasks. This integration could be accomplished by including STRING in eSPERTo's pipeline, or/and by deriving new resources compatible with NooJ based on STRING's own resources.

In the future, we also need to study different configurations of the resources to identify the parameterization that will produce the best paraphrases for each application scenario.

Acknowledgments. This research was supported by Fundação para a Ciência e Tecnologia (FCT), under exploratory project eSPERTo (Ref. EXPL/MHC-LIN/2260/2013). Anabela Barreiro was also funded by FCT through post-doctoral grant SFRH/BPD/91446/2012. The authors would like to thank Max Silberztein for his prompt support and guidance with all matters related to NooJ.

References

1. Amaral, R., Meinedo, H., Caseiro, D., Trancoso, I., Neto, J.: A prototype system for selective dissemination of broadcast news in European Portuguese. EURASIP J. Adv. Sig. Process. **2007**(1), 1–11 (2007)
2. Barreiro, A.: Port4NooJ: Portuguese linguistic module and bilingual resources for machine translation. In: Blanco, X., Silberztein, M. (eds.) Proceedings of the 2007 International NooJ Conference, pp. 19–47. Cambridge Scholars Publishing, Newcastle (2008)
3. Barreiro, A., Batista, F., Ribeiro, R., Moniz, H., Trancoso, I.: OpenLogos semantico-syntactic knowledge-rich bilingual dictionaries. In: Calzolari, N., Choukri, K., Declerck, T., Loftsson, H., Maegaard, B., Mariani, J., Moreno, A., Odijk, J., Piperidis, S. (eds.) Proceedings of the Ninth International Conference on Language Resources and Evaluation (LREC 2014). ELRA, Paris (2014)

4. Barzilay, R., McKeown, K.R.: Sentence fusion for multidocument news summarization. Computational Linguistics **31**(3), 297–328 (2005)
5. Fialho, P., Coheur, L., Curto, S., Cláudio, P.: Meet EDGAR, a tutoring agent at Monserrate. In: Proceedings of the 51st Annual Meeting of the ACL. Omnipress, Madison (2013)
6. Lin, C.Y.: ROUGE: a package for automatic evaluation of summaries. In: Text summarization branches out. Proceedings of the ACL-04 workshop, p. 74. ACL, Stroudsburg (2004)
7. Mamede, N.J., Baptista, J., Diniz, C., Cabarrão, V.: String: an hybrid statistical and rule-based natural language processing chain for Portuguese. In: Caseli, H., Villavicencio, A., Teixeira, A., Perdigao, F. (eds.) Proceedings of the 10th International Conference on Computational Processing of the Portuguese Language. Series Demo Session, PROPOR 2012. Springer, Heidelberg (2012)
8. Mota, C., Carvalho, P., Barreiro, A.: Port4NooJ v3.0: integrated linguistic resources for Portuguese. In: Calzolari, N., Choukri, K., Declerck, T., Goggi, S., Grobelnik, M., Maegaard, B., Mariani, J., Mazo, H., Moreno, A., Odijk, J., Piperidis, S. (eds.) Proceedings of LREC 2016. LREC, Portoroz (2016)
9. Mota, C., Carvalho, P., Raposo, F., Barreiro, A.: Generating paraphrases of human intransitive adjective constructions with Port4NooJ. In: Okrut, T., Hetsevich, Y., Silberztein, M., Stanislavenka, H. (eds.) NooJ 2015. CCIS, vol. 607, pp. 107–122. Springer, Heidelberg (2016). doi:10.1007/978-3-319-42471-2_10
10. Neto, J.P., Meinedo, H., Amaral, R., Trancoso, I.: A system for selective dissemination of multimedia information resulting from the ALERT project. In: 2003 ISCA Workshop on Multilingual Spoken Document Retrieval (MSDR2003). ISCA, Hong Kong (2003). http://www.isca-speech.org/archive_open/msdr
11. Pardo, T., Rino, L.: TeMario: a corpus for automatic text summarization. Technical report, NILC-TR-03-09, Sao Paulo (2003)
12. Ribeiro, R.: Summarizing spoken documents - avoiding distracting content. Ph.D. thesis. Instituto Superior Técnico, Lisboa (2011)
13. Ribeiro, R., Martins de Matos, D.: Centrality-as-relevance: support sets and similarity as geometric proximity. J. Artif. Intell. Res. **42**, 275–308 (2011)
14. Rinaldi, F., Dowdall, J., Kaljurand, K., Hess, M., Mollá, D.: Exploiting paraphrases in a question answering system. In: ACL-2003, Second International Workshop on Paraphrasing: Paraphrase Acquisition and Applications, pp. 25–32. ACL, Sapporo (2003)
15. Scott, B.B.: The logos model: an historical perspective. Mach. Transl. **18**(1), 1–72 (2003)
16. Silberztein, M.: Formalizing Natural Languages: The NooJ Approach. Wiley, London (2016)
17. Trancoso, I., Neto, J.P., Meinedo, H., Amaral, R.: Evaluation of an alert system for selective dissemination of broadcast news. In: INTERSPEECH - Proceedings of the 8th European Conference on Speech Communication and Technology (EUROSPEECH-INTERSPEECH 2003), pp. 1257–1260. ISCA Archive (2003). http://www.isca-speech.org/archive/eurospeech

Semantic Analysis and Its Applications

Endpoint for Semantic Knowledge (ESK)

Maria Pia di Buono[✉]

Faculty of Electrical Engineering and Computing,
University of Zagreb, Zagreb, Croatia
mariapia.dibuono@fer.hr

Abstract. This work arises from the evaluation of the existing methods used to process knowledge stored in on-line repositories and databases.

Such methods represent an attempt to improve the techniques of Knowledge Extraction (KE) and Representation (KR) to guarantee a more accurate answer to users' information request.

Different research groups are nowadays developing various endpoint systems, as for instance Virtuoso, which is devoted to run queries against online Knowledge Bases (KBs), mainly DBpedia.

We present an Endpoint for Semantic Knowledge (ESK), a system which integrates NooJ Linguistic Resources (LRs) in an environment suitable for a semantic search engine. ESK is structured as a SPARQL endpoint, which applies a deep semantic analysis, based on the development of a matching process between a set of machine semantic formalisms and a set of Natural Language (NL) sentences.

Keywords: DBpedia · ESK · SPARQL · CIDOC CRM · NooJ Linguistic Resources

1 Introduction

The treatment of knowledge[1] by means of machines is strictly linked with two procedures, generally considered, and handled, as separate steps: the Knowledge Representation (KR) and Knowledge Extraction (KE) tasks. These tasks require the development of a formal semantic description which allows to map natural languages with a machine-readable formal representation. The possibility of mapping natural languages with machine formalisms entails the development of adequate descriptive models.

Such descriptive models represent the way in which we formalize human knowledge into machine formalisms. In fact, if knowledge processing requires processing of its constituent elements, expressed in natural languages, and furthermore if human knowledge is not machine-readable, then we have to apply are presentation model suitable to 'convert' natural-language elements into a machine-readable format.

---

[1] This work was carried out while the author was affiliated with the Dept. of Political, Social and Communication Sciences - University of Salerno, Fisciano (SA), Italy.

© Springer International Publishing AG 2016
L. Barone et al. (Eds.): NooJ 2016, CCIS 667, pp. 223–233, 2016.
DOI: 10.1007/978-3-319-55002-2_19

For these reasons, addressing the problem of knowledge treatment by means of machines may be achieved dealing with knowledge as far as both its representation and extraction are concerned.

In this paper, we introduce a methodology that foresees the interaction of NooJ [1] Linguistic Resources (LRs) and machine formalisms to efficiently handle knowledge. Using NooJ in order to develop our LRs allow us to apply two elements:

– A rational grammar, which is also out of context and describes a language larger than contextual ones;
– A series of constraints, which exclude certain sequences recognized by pure contextual grammars keep only those sequences that truly belong to the desired contextual language [2].[2]

Thus, our methodology applies a representation model which defines rules and constraints for developing NooJ LRs, suitable to achieve a deep semantic analysis.

Furthermore, the proposed model integrates formal-semantic annotations into NooJ LRs in order to guarantee a fine-grained linguistic analysis developing an effective methodology that enables a consistent and meaningful Knowledge Processing (KP).

2   State of Art

Different research groups are focused on the development of models suitable to process knowledge stored in Knowledge Bases (KBs), mainly by means of Ontology Learning (OL) approaches.

OL represents the process through which we extract conceptual knowledge and elements from different inputs, in order to build an ontology. Several methods are applied in OL process, such as Machine Learning (ML), Knowledge Acquisition (KA), Natural Language Processing (NLP), Information Retrieval (IR), Artificial Intelligence (AI), reasoning and database management.

In order to extract concepts from texts and provide inferences on ontological knowledge, most researches are focused on different learning processes for populating ontology. Supporting the construction of ontologies and populating them with instantiations of both concepts and relations is commonly referred to as onto OL [3].

OL may be achieved through a manual development or a (semi-)automatic procedure. Manual ontology acquisition is basically used by knowledge engineering or domain experts, and it includes tools such as Protege-2000 [4, 5] and OntoEdit [6].

Among (semi-)automatic procedures, the Linked Data Mining approach seems to be a promising way to achieve the KP tasks. It intends to detect meaningful patterns inside Resource Description Framework (RDF) graphs, via statistical schema induction [7–9] or statistical relational learning methods. This research area often applies clustering approaches in order to group interconnected resources.

---

[2] Translation by the editor. (…) – une grammaire G rationnelle ou hors contexte qui définit un langage plus grand que le langage contextual qu'on veut décrire – une série de contraintes qui excluent certaines sequences reconnues par G pur ne garder que les séquences qui appartiennent véritablement au language contextual voulu [2].

New research trends aim at developing efficient endpoint systems,[3] as for instance Virtuoso,[4] which is devoted to run queries against online KBs, mainly DBpedia.[5] Virtuoso offers a SPARQL Service Endpoint structured as a Server-hybrid Web Application, namely a Universal Server, which provides SQL, XML, and RDF data management in a single multithreaded server process.

Generally speaking, such endpoints allow to access KBs by means of an interface from which it is possible to run SPARQL queries, as it happens with query editors. In other words, endpoints do not process Natural Language (NL) queries, but require a structured query, which means that they handle just the information stored as RDF triples into KBs.

3   Methodology

Our methodology relies heavily on a linguistic processing phase, and requires the development of robust LRs.

In order to achieve the linguistic processing, we apply a model suitable to describe linguistic phenomena and formalisms in a coherent and consistent way.

Such a formal linguistic description represents the basis on which we develop a semantic representation of a specific knowledge domain.

In other words, our approach aims at creating a linguistic description of distributional and transformational properties, which means a Lexicon-Grammar (LG) of simple sentences[6] [10], suitable to describe concepts in a knowledge domain.

Thus, our methodology is based on analyzing and developing particular grammars, not abstract constraints on whole classes of grammars [11].

---

[3] As reported in the Web site for the Semantic Web, "A SPARQL endpoint enables users (human or other) to query a knowledge base via the SPARQL language. Results are typically returned in one or more machine-processable formats. Therefore, a SPARQL endpoint is mostly conceived as a machine-friendly interface towards a knowledge base. Both the formulation of the queries and the human-readable presentation of the results should typically be implemented by the calling software, and not be done manually by human users". Source: http://semanticweb.org/wiki/SPARQL_endpoint.html.

[4] http://dbpedia.org/sparql.

[5] Other samples of endpoints, which allow to access DBpedia KB, are indicated at http://www.w3.org/wiki/SparqlEndpoints. Anyway, due to the spread of several KBs, different endpoints are available. For more information on current alive SPARQL endpoints, see: http://www.w3.org/wiki/SparqlEndpoints.

[6] Indeed, as [9] states, LG considers the simple sentence as the base unit of analysis, which means for LG: *A lexicon-grammar is constituted of the elementary sentences of a language. Instead of considering words as basic syntactic units to which grammatical information is attached, we use simple sentences (subject-verb-objects) as dictionary entries. Hence, a full dictionary item is a simple sentence with a description of the corresponding distributional and transformational properties.*

These grammars describe specific linguistic behaviors, also considering the verb valency and verb centrality[7] [12]. In other words, the verb phrase (VP) may be used to define semantic-role sets which represent concepts.

These semantic-role sets are developed applying principles of dependency, semantic expansion and informational conditions [13].

Thus, we take into account that certain words depend on the presence of other words to form an utterance.

Furthermore, some combinations of words and their dependents are more likely than others, which means there exists a likelihood principle.

Finally, we also consider the reduction principle, which means that words in high likelihood combinations can be reduced to shorter forms, and sometimes omitted completely.

This theoretical framework allows a deep linguistic analysis, which performs both object/term and synonym identification, and also recognizes relations which describe concepts. Since it is based on a deep analysis and formalization of linguistic phenomena, our approach can also ensure portability to other knowledge domains, preserving consistency and disambiguation.

3.1   Formal Linguistic Model

Dealing with concepts requires a deep linguistic analysis, due to the combinatorial features of natural languages, and their internal intricacy. The process of extracting such kind of data entails several steps, aimed at converting texts into a machine-readable format and at identifying the information to be extracted.

At present, ontology and RDF-based methods are the most promising solutions for conceptual description and modelling of textual data. Indeed, RDF, ontologies, and NLs share some characteristics, which makes possible to assume a correspondence among these formalisms. In our opinion, such correspondence is retrievable into RDF predicates, ontological properties, and VPs (i.e., the syntactic behaviors of semantic predicates).

Actually, in all of these three formal descriptions, RDF, ontology and natural language, two atomic elements are connected by a central element. Such central element stands for a trigger able to attract the other two elements, which means capable to establish a relationship between them (Fig. 1).

Therefore, these three formal descriptions present a similar logic structure, on the basis of which we may suppose the existence of a parallelism among them. In other words, in such formalisms, we may assume the presence of sets which hold the trigger and two elements involved in the relationship.

---

[7] Sentence is considered as an ensemble organisé (organized set) in which mots (words) are constituants (constituent elements). A word in a sentence is not isolated as in the dictionary, due to the fact that we can perceive connexions (connections) between such word and its neighbors.

Thus, on the basis of such formal model we structured exhaustive and descriptively taxonomic and ontological LRs (i.e., electronic dictionaries, syntactic matrix tables and local grammars), using NooJ [14].

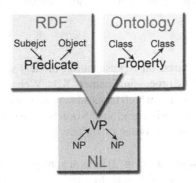

Fig. 1. Schema of formal descriptions for RDF, ontology and NL

4   Endpoint for Semantic Knowledge

This methodology has been tested in ESK, a system which integrates NooJ LRs in an environment suitable for creating a semantic search engine. ESK is structured as a SPARQL endpoint, which applies a deep semantic analysis, based on the development of a matching process between a set of machine semantic formalisms and a set of NL sentences.

ESK proposes an interface in which users may insert NL queries and run them again a KB. Furthermore, ESK is set to offer a tool for on-line processing of unstructured texts and Web pages, in order to generate semantically tagged documents. Such a system has to be based on a deep linguistic analysis, which aims at processing both NL and structured information.

Actually, ESK manages three kinds of inputs, which refer to three independent resources and produce three different outputs:

1. Users' NL query, which refers to Linked Open Data (LOD) KBs, such DBpedia or Europeana repository.
2. URL(s), inserted by users, which relates to Web pages and accomplishes a text retrieval.
3. Unstructured texts, uploaded by users, which concern full-text analysis and produce tagged texts.

ESK is just a beta-version endpoint, built to cope with the Archaeological knowledge domain, and it is set up for Italian. Therefore, the LRs used during the linguistic processing and the source KBs are limited in size and number. This means that, concerning LRs, ESK applies the Italian module of NooJ [15], plus the Archaeological Italian Electronic Dictionary (AIED) and a series of related Finite-State

Automata/Transducers (FSA/FSTs) [16]. On the other hand, concerning KBs, ESK is structured to run a query against Europeana KBs and DBpedia.

ESK is implemented by dint of PHP code, a server-side scripting language developed for Web application, which is suitable for providing dynamic, user-oriented contents.

4.1 Workflow

The system workflow is based on representation models applied to all LRs, which represent objects of linguistic processing, namely KBs, Web pages and full texts. Therefore, we develop an architecture, which takes advantage from the semantic information stored in LRs and is based on the integration of NooJ. Such system architecture integrates NooJ into a Web application in order to (re)use the representation models used to formalize LRs.

Hereby, ESK system workflow (Fig. 2) aims at applying a linguistic analysis to three type of inputs, integrating semantic annotation tasks for each one of these.

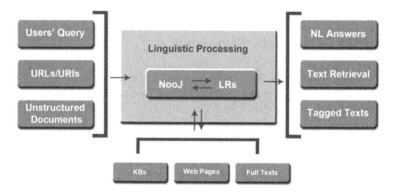

Fig. 2. ESK workflow

The system architecture is structured as follows:

- Users insert input into ESK web form, according to their information needs, which means that they may insert (I) a query, structured as simple sentence; (II) a URL/URI, indicating a Web page to be processed; (III) a text file which they intend to analyze and tag.
- ESK records inputs into text files and applies a normalization process, i.e., it deletes punctuation marks, and so on.
- ESK handles such text files as inputs for NooJ, and it also manages LRs which have to be associated to the running linguistic analysis.
- NooJ performs linguistic processing, which is different for each of the three inputs. As we will see, this means that ESK applies different LRs, namely FSA/FSTs, in order to return different outputs.

ESK may return the following different results:

- An NL answer to users' query. This means a list of results as literals, which cope with the request and references of resources for these results.
- A Text Retrieval, which is intended as (a set of) semantically annotated strings, referring to Atomic Linguistic Units (ALUs) and concepts, extracted from the page indicated by users' URL.
- A Tagged Text. This means a document annotated in XML and also by metadata schemata, e.g., RDF, Europeana Data Model (EDM) and Simple Knowledge Organization System (SKOS).[8]

Additionally, ESK may also handle the results of the statistical analysis performed by means of NooJ, e.g., frequencies, standard score, and so on.

It is worth noting that NooJ allows to analyze also selected parts of corpora, thus we may apply statistical measures to only these parts, so obtaining focused measures.

5 Linguistic Processing

The linguistic processing phase, which is managed through the linguistic engine of NooJ, is focused on various NLP tasks, as they follow:

- Named Entity Recognition and Classification;
- Taxonomic relation extraction;
- Property extraction.

These tasks represent the main goals in KP, due to the fact that they allow to recognize and extract specific concepts used in a certain domain, describing the existing relationships among them.

According to our theoretical and practical framework, such LRs are developed applying three types of rules, as they follow:

- Selection restriction and co-occurrence rules (based on an accurate lexicon formalization).
- Taxonomic rules (derived from taxonomy prescriptions of Italian Institute for Catalogue and Documentation);
- Semantic rules (referred to CIDOC Conceptual Reference Model - CRM) [17].

In other words, our methodology allows to define a set of rules and constraints suitable to process three different type of inputs in ESK, improving both the KR and the KE.

5.1 Natural Language Queries

In order to process this fist type of input, the main task to accomplish is the processing of user's query by means of a linguistic analysis. Such an analysis aims at annotating

---

[8] For more information, see: http://en.wikipedia.org/wiki/Europeana and http://en.wikipedia.org/wiki/Simple_Knowledge_Organization_System.

the query using a domain-independent semantic data model (i.e., DBpedia cross-domain ontology).

Thus, ESK records the query on a text file and set it as a NooJ parameter, namely as a variable used as one of the inputs to the subroutine.

During this analysis, other variables are assigned to NooJ subroutine, namely the AIED and a set of grammars in the form of FSA [18].

Therefore, if the input is a NL question, after the linguistic processing phase, ESK runs the corresponding query against a remote endpoint proceeding as it follows:

- Open and access to the endpoint;
- Execute the query;
- Display the results.

5.2   URL/URI Web Tool

In presence of a specific user's query, ESK connects to a Web page and save its contents into a text file, applies to it a normalization process and consequently set it as a NooJ variable. During the linguistic processing of such kind of inputs, we apply the AIED to recognize ALUs/entities and associate Linguistic LOD references to each one of them.

Actually, this second subtask, namely data representation, involves appropriate operations on the RDF-based data layer, which includes the mapping of OWL concepts to object-oriented classes with methods for interrelations and domain-specific rules, used to generate and consolidate all processes (e.g. CIDOC CRM ontology).

Such process of data representation aims at analyzing the information stored in Web documents, which means that we may directly retrieve information from any URLs.

According to our formal model, electronic dictionaries entries (simple words and ALUs) are the subject and the object of the RDF triples which are traceable inside sentence structures.

In addition, as regards declarative sentences, RDF gives the possibility to recognize sentences conveying information of the type "X is an element of Y", which also is a type of recursive and iterative (therefore productive) structure.

Enclosing information and constraints, derived from the RDF data representation and the CIDOC CRM, inside FSA allows to identify and extract entities and properties.

In other words, we retrieve and extract ALUs/entities and VPs/properties from texts using a formalization based on a match among elements in nuclear sentence and the domain-specific ontology.

Therefore, ESK results for a URL-based input is represented by a list of elements, namely entities, retrieved from the queried Web page, and structured as follows:

```
<E21><Person>Peter Chad Tigar Levi</Person></E21> (…)
<P14></P14><E50><Date>16 May 1931 - 1 February 2000</
Date></E50>
<P98>was</P98> (…)
```

```
<E7><Activity1>archaeologist</Activity1> (...)
<Activity2>travel writer</Activity2></E7>⁹
```

[Example of a tagged text extracted from a URL-based input]

Furthermore, considering that users may insert also a URI in the input field, we develop FSA suitable to retrieve this structured information. In such kind of FSA, we use the nodes in order to recognize labels used inside RDF documents, which are stored, for example, in DBpedia KB.

This means that:

- First, we process tags which describe elements semantically.
- Subsequently, we analyze which values are assumed for such descriptions (literal or numeric).

For this kind of input, ESK returns result in the form of structured information stored in the URI used.

5.3 Unstructured Texts

ESK also offers a function suitable for processing a text file uploaded by users, applying Archaeological LRs[10] [17]. Actually, during this phase, using FSA as input variables, NooJ allows to convert an unstructured text into a document formalized according to a specific-domain data model.

Furthermore, ESK guarantees the possibility to export such results, using RDF and SKOS, and also to use LLOD URIs to tag the AIED entries.

This procedure for unstructured-text processing is accomplished by means of the following steps:

- NooJ processes a text, parses it, and locates all the terminological ALUs inside the given text;
- If retrieved ALUs belong to Agent or Place classes, we associate them to URIs in order to integrate LLOD resources.
- Subsequently, the ALUs retrieved are conceptually described by means of SKOS/RDF schemata and features, as for instance those used in EDM;
- At the same time, RDF triples are transformed into EDM tags in which concepts and relationships, for instance E21 or P14, are rewritten by means of corresponding "edm:Agent" or "edm:begin";
- Finally, NooJ output is transformed into a full EDM XML Schema.

Therefore, ESK converts this unstructured text into a string structured according to EDM XML Schema, as in the result which follows:

---

⁹ Result derived from the analysis of Wikipedia entry "Peter Levi". https://en.wikipedia.org/wiki/Peter_Levi.

[10] This LRs are based on the Italian Linguistic Module, created by [14] and maintained by the team of the Laboratory of Computational Linguistics "Maurice Gross" of University of Salerno.

```
<edm:Agent>Num</edm:Agent>
<edm:begin>born in</edm:begin>
<edm:TimeSpan>Date</edm:TimeSpan>[11]
```

[Example of output from an unstructured-text input]

It is worth to remember that such a result comes from the ontological and semantic constraints which are applied during the formalization process. In other words, the VP requires the co-occurrence of an Num as N0 and a Date as N1, due to the semantic behavior assumed by the verb.

6   Conclusions and Future Works

In this paper, we proposed a methodology which aims at creating a framework for KP, based on a deep linguistic analysis.

Also, we propose an architecture, which takes advantage from semantic information stored both in NooJ electronic dictionaries and FSA/FSTs. Furthermore, this architecture may also map linguistic tags (i.e., POS) and structures (i.e., sentences, ALUs) to domain concepts employing metadata from conceptual schemata.

It is worth noticing that our approach allows to devise efficient strategies for representing deep attributes and semantic properties of natural languages also concerning machine formalisms.

In order to improve our methodology the short-term goal is to further investigate formalization of natural languages from a LG perspective, particularly with respect to domain-specific linguistic features and machine-language equivalences.

The long-term goals may be identified in the integration of our method with a hybrid approach to KP, in order to achieve high quality knowledge representation and extraction by combining probabilistic and linguistic information.

Furthermore, we intend to integrate an index-data structuring and a query evaluation process into our environment and test the system in a consistent way, on other KBs, in order to propose an independent-domain approach.

References

1. Silberztein, M.: The NooJ manual (2003). http://www.nooj4nlp.net/pages/references.html
2. Silberztein, M.: La formalisation des langues. L'approche de NooJ. ISTE, London (2015)
3. Lehmann, J., Völker, J.: An introduction to ontology learning. In: Lehmann, J., Völker, J. (eds.) Perspectives on Ontology Learning, pp. ix–xvi. AKA/IOS Press (2014)
4. Grosso, W.E., Eriksson, H., Fergerson, R.W., Gennari, J.H., Tu, S.W., Musen, M.A.: Knowledge modeling at the millennium. In: Gaines, B.R., Kremer, R.C., Musen, M. (eds.) Proceedings of the 12th Workshop on Knowledge Acquisition, Modeling and Managmenet (KAW 1999), Banff, Canada, pp. 1–36 (1999)

---

[11] It is worth noticing that in such example we do not use *edm:year* in order to tag birth/death date, due to the fact that such property refers to an event in the life of the original analogue or born digital object. Therefore, *edm:year* property is not applicable to the class Agent.

5. Noy, N.F., Fergerson, Ray, W., Musen, Mark, A.: The knowledge model of Protégé-2000: combining interoperability and flexibility. In: Dieng, R., Corby, O. (eds.) EKAW 2000. LNCS (LNAI), vol. 1937, pp. 17–32. Springer, Heidelberg (2000). doi:10.1007/3-540-39967-4_2

6. Sure, Y., Erdmann, M., Angele, J., Staab, S., Studer, R., Wenke, D.: OntoEdit: collaborative ontology development for the semantic web. In: Horrocks, I., Hendler, J. (eds.) ISWC 2002. LNCS, vol. 2342, pp. 221–235. Springer, Heidelberg (2002). doi:10.1007/3-540-48005-6_18

7. Bühmann, L., Lehmann, J.: Universal OWL Axiom Enrichment for Large Knowledge Bases. In: Teije, A., Völker, J., Handschuh, S., Stuckenschmidt, H., d'Acquin, M., Nikolov, A., Aussenac-Gilles, N., Hernandez, N. (eds.) EKAW 2012. LNCS (LNAI), vol. 7603, pp. 57–71. Springer, Heidelberg (2012). doi:10.1007/978-3-642-33876-2_8

8. Bühmann, L., Lehmann, J.: Pattern based knowledge base enrichment. In: Alani, H., et al. (eds.) ISWC 2013. LNCS, vol. 8218, pp. 33–48. Springer, Heidelberg (2013). doi: 10.1007/978-3-642-41335-3_3

9. Völker, J., Niepert, M.: Statistical schema induction. In: Antoniou, G., Grobelnik, M., Simperl, E., Parsia, B., Plexousakis, D., Leenheer, P., Pan, J. (eds.) ESWC 2011. LNCS, vol. 6643, pp. 124–138. Springer, Heidelberg (2011). doi:10.1007/978-3-642-21034-1_9

10. Gross, M.: Lexicon-grammar and the syntactic analysis of French. In ACL (eds.) Proceedings of the 10th International Conference on Computational Linguistics, pp. 275–282. ACL, Stroudsburg (1984)

11. Gross, M.: On the failure of generative grammar. Language 55(4), 859–885 (1979). The Linguistic Society of America

12. Tesnière, L.: Eléments de syntaxe structurale. Klincksieck, Paris (1959)

13. Harris, Z.S.: Language and Information. Columbia University Press, New York (1988)

14. di Buono, M.P.: Information extraction for ontology population tasks. An application to the italian archaeological domain. Int. J. Comput. Sci. Theor. Appl. 3(2), 40–50 (2015). ORB Academic Publisher

15. Vietri, S.: The Italian module for NooJ. In: Basili, R., Lenci, A., Magnini, B. (eds.) Proceedings of the First Italian Conference on Computational Linguistics, CLiC-it 2014, pp. 389–393. Pisa University Press, Pisa (2014)

16. di Buono, M.P., Monteleone, M., Elia, A.: Terminology and knowledge representation italian linguistic resources for the archaeological domain. In: Proceedings of 25th International Conference on Computational Linguistics (COLING 2014) - Workshop on Lexical and Grammatical Resources for Language Processing (LG-LP 2014), pp. 24–29. ACL, ACL Web Anthology (2014). www.aclweb.org/anthology/W14-5804

17. Doerr, M.: The CIDOC conceptual reference module: an ontological approach to semantic interoperability of metadata. AI Mag. 24(3), 75 (2003)

18. di Buono, M.P.: Semi-automatic indexing and parsing information on the web with NooJ. In: Okrut, T., Hetsevich, Y., Silberztein, M., Stanislavenka, H. (eds.) NooJ 2015. CCIS, vol. 607, pp. 151–161. Springer, Heidelberg (2016). doi:10.1007/978-3-319-42471-2_13

Using Text Mining and Natural Language Processing to Support Business Decision: Towards a NooJ Application

Francesca Esposito[1]([✉]) and Maddalena della Volpe[2]([✉])

[1] Department of Political, Social and Communication Sciences,
University of Salerno, Salerno, Italy
fraesposito@unisa.it
[2] University of Suor Orsola Benincasa, Naples, Italy
maddalena.dellavolpe@unisob.na.it

Abstract. Decision-making process has become extremely difficult especially for the large amount of textual data that companies must analyse to be competitive. The use of Natural Language Processing and Text mining in data discovery allows extracting knowledge from business texts that in the majority occur in unstructured form. The Decision Support System and the Information Technology departments face the new challenges that change poses, relying on linguistic analysis capabilities, no longer based on keyword research but on the syntactic properties, lexical and semantic word. In this paper, we focused on document-driven decision support, describing ways in which business communication performance can be improved by using a natural language interface as NooJ. In order to achieve our goals, we developed Linguistic Resources typically used in Economy knowledge domain, with regard to compound words and multi-word atomic linguistic units (MWALUs).

Keywords: Business decision · Document-driven DSS · Natural language processing · Text mining · NooJ application

1 Introduction

A Decision Support Systems (DSS) is typically a computer program application that analyses business data and presents it so that managers can make business decisions more easily. This application finds its roots in the Information Retrieval (IR) field and may present information graphically including an expert system or Artificial Intelligence (AI). The continuous changes and the exponential growth of the amount of data that the companies have to analyse in order to compete in the market, impose them the need to develop DSSs, integrating new functions and new applications.

Innovation challenges is disregarding the traditional functions of information management, focused mainly on analysis of internal data storage: companies must consider many variables and sources.

Even if companies recognize that new information systems are required, in practice this has not yet been applied. The continuous flow of data, produced by enterprises,

© Springer International Publishing AG 2016
L. Barone et al. (Eds.): NooJ 2016, CCIS 667, pp. 234–245, 2016.
DOI: 10.1007/978-3-319-55002-2_20

requires a new sampling method to manage information: an appropriate way to analyse and collect each single datum regarding its features.

This strategic orientation is much easier to apply into startup companies while it requires a profound change in the mentality of Information Technology (IT) department of big company activities. The learning culture and the willingness to revise his opinions on management systems is difficult in certain organizations.

Thus, a change must concern the greater propensity to the discovery of data: especially data discovery and extraction. Through Natural Language Processing (NLP) and Text Mining (TM), we will frequently be able to extract patterns, resources and opportunities as old and new data sources. In addition, the introduction of deep linguistic analysis of business documentation in DSS would allow:

- The reduction of time spent on measurement and analysis of the project;
- The possibility of a massive control of the documentation;
- The non-dispersion of information related to the company;
- The immediate identification of benchmarks;
- The immediate identification of processes but also absent;
- The systematic support to the improvement of company documents;
- The constant monitoring on the quality and adequacy of the activities.

In this paper, we focused firstly on the relation between DSS and linguistic analysis. We dealt with document-driven DSS approach: analysing text or business document as information at the bottom of decisions using NLP and TM technologies. To realise this approach with NooJ, we proposed a model of document-driven analysis underlining the role of three elements that influence a business fact-finding survey.

A combined approach is adopted for carrying out a large-coverage of Italian Linguistic Resources and at the same time discovering new Linguistic Resources, recognising properties useful to the creation of automata and upgrading of electronic dictionaries.

2   Supporting Business Decision with Documents Analysis

Decision Support Systems (DSS) are generally defined as a collection of computerized data system that supports some decision-making actions [1]. In the 1960s, researchers began systematically studying the use of computerized quantitative models to assist in decision-making and planning, at any level of organization. Sprague and Carlson, quoted by Power [2], defined DSS as "a class of information system that draws on transaction processing systems and interacts with the other parts of the overall information system to support the decision-making activities of managers and other knowledge workers in organizations" [3].

In the last decade, the technology upgrade created new computerized decision support applications in many disciplines. They consisted of interactively computer-based systems that emphasised manipulating quantitative models, accessing and analysing large databases, and impacted on decision-making structures. Applications comprised of following types: model-driven DSS, data-driven DSS, communication-driven DSS, document-driven DSS, knowledge-driven DSS and web-driven DSS.

These different models are not mutually exclusive, although they are used depending on the needs of enterprises. The DSS are generally capable of collecting and representing several type of information, such as comparative data, accessing information assets that include relational and legacy data sources, providing past experiences as well as projecting future choices according to assumptions or new data, consequences of decisions.

This turns out to be necessary for managers and companies that are transforming the way in which they are designing and constructing new information management capabilities. Companies are increasingly looking for new data to be used as strategic resource, developing their analysis capabilities and refining tools that support concretely and help top managers to take better decisions.

Decision-making is the process of developing and analysing alternatives to make a decision, a choice from the available alternatives [4]. Most decisions are made to solve problems or generally improve business performance. Making decisions is now even more difficult for companies, which have to deal with the complexity of the business system: taking decisions means doing in-depth analysis of each asset and considering the rapid change of business equilibria. Managers need to make the best decisions in the shortest possible time and at the lowest cost.

In addition to this, companies have to analyse a huge amount of data that often appear unstructured, not categorized in traditional data warehouses. It is the case of business documents: companies use them to describe their activities, characteristics, or strategies and they could be divided into two categories: oral and written. Oral documents consist in conversations transcribed while written documents can be written reports, strategic plans, catalogues, memos and even e-mail. These unstructured forms of knowledge require the use of technologies that are able to recognize the hidden meaning in the texts. Natural Language Processing (NLP) and Text Mining (TM) can meet these needs: Semantic Technologies market is experiencing an increase and the creation of companies that offer software solutions in this regard.

Document-Driven DSS is a relatively new field in Decision Support and it is focused on the information retrieval and management of unstructured documents. Particularly, Decision Support driven by documents aim to manage, retrieve and manipulate unstructured information in a variety of electronic formats [1]. It assists manager and IT unit of companies in knowledge categorization, deployment, inquiry, discovery and improving communication.

Back in 1996, Fedorowicz [5] estimated that American businesses store almost 1.3 trillion documents and only 5 to 10% of these documents are available to managers for use in decision-making just because they were not standardized in a uniform pattern or structure. Managers need a way to transform these documents into usable formats that can be matched and processed to support decision-making [6]. Moreover, in 1998 Merrill Lynch sustained that somewhere around 80–90% of all potentially usable business information may originate in unstructured form. This rule is not demonstrated scientifically but expresses a principle, often inferred from experience, that it is indicated as valid in most cases.

Increasing contents recovered by Web, the unstructured data, such as text, will become the predominant data type stored online. The data grew too fast and exceeded human capacity to retrieve and utilize. As mentioned, managerial decision-making process can be highly dependent on hidden information in text documents. However, careful reading and sorting of large amount of documents is A time-consuming work. This type of activity wastes working time of managers and in the end, it can even cause wrong decisions. Companies could gather innovation challenges using this type of technologies that combine different textual data and create a new generation of business functions with the support of specific knowledge.

Kopackova, Komarkova and Sedlak [7] noticed that TM could be used for preprocessing of textual information in order to find hidden knowledge and ease the process of decision-making. TM and NLP as technologies have their roots in linguistics but in recent years, they fulfilled a pregnant role in Analytics market, moreover for analysis interfaces of term-extractions. The ability to recognize linguistic features in text, even concepts and sentiments, and to extract them to databases, is now an important feature for these tools and an opportunity for companies. The collaboration between Machine Learning and linguistic tools, that support functions such as semantic disambiguation, is the only way to transform texts into manageable knowledge [8].

Retrieved data through keyword search is no longer enough. Early approaches of TM would treat a text source as a set of words. They evolved, starting from basic linguistic forms up to recognize more complex forms, as Multi-Word Atomic Linguistic Units (MWALUs). Basic lexical analysis might count frequencies of words in order to classify documents by topic. Although, there was no ability to understand the semantic aspect, the meaning contained in a document. The complexity of the texts and the potential knowledge hiding require a much more comprehensive study of the linguistic properties, the study of terminology and the resulting semantic annotation: it describes the use of language and connects the various terms in specific meanings, as we will see subsequently with the semantic expansion concept.

3 Enterprise-Data-Technology Document-Driven DSS

As we have seen, NLP and TM software could be integrate in the DSS for business retrieving information by documents analysis. The choice of texts to be analysed may fall both within the company and externally. Many companies offering semantic technology solutions, which are focused on the analysis of external data, as, can be conversations on the web, opinions and data about economic trends. However, we have to consider the role of internal information. They usually appear in unstructured form, but contain a huge potential: they belong to the company and have already targeted on it, and should not be identified in the sea of information that nowadays crowd the Web. The collaboration between internal and external data allow the development of a series of tools and multidimensional analysis models.

Below we identified three element that we have to consider if we introduce a NLP or TM software in DSS of a company: enterprise, data and technology.

Enterprise. Traditional analytics should consider the whole enterprise, trying to understand how the sharing of technology and human resources data across all organizational units will lead to achieving the general analytical goals. It is necessary for innovative startup and early adopter companies to coordinate data collection at all levels of the company [9]. In large organizations, it is necessary to introduce the analysis of data to assist in the various departments: marketing, finance, product development, strategy and information systems.

Establishing where to place this type of Technology within the enterprise depends on the goals. An aspect not to be underestimated is the human one, which must always be more specialized in data management and transfer knowledge: it is from the combination of man and machine that we can get the best results.

Furthermore, each sector has a different way to communicate their own business. For example, if we analyse a Business Plan written by an Agri-food company, we may find in the text many words referring not only to Agriculture knowledge field, but also to others as Biology, Chemistry, Botany, Ecology and more. Therefore, it is necessary knowing the company features, its market, its products, its development phase.

Data. Every day around the world millions digital data are created. Big Data processing requires considerable computing power, appropriate technology and well-defined resources, which go beyond current management systems and data storage capabilities. These data have aroused a great interest from the academic and business world: an impetuous demand for services has generated an uncontrollable range of analytical Big Data solutions. This market was characterized in 2014 by an increase of +25%, due not just to the maturity of the analytical tools used as a storage services and low-cost clustering that have attracted the interest of small businesses and large enterprises. Big Data are high-volume, high-velocity and/or high-variety information assets that demand cost-effective, innovative forms of information processing enabling enhanced insight, decision-making, and process automation [10].

There are two categories of textual data: unstructured or structured. Structured data are the typical elements stored in a database, carefully selected and categorized. These data are usually the result of years of work, in which they are collected, sorted and stored. Unstructured data usually refer to data that do not reside in a traditional row and column database. For example, unstructured textual data include text and multimedia content such as emails, documents and more. This type of data is more difficult to analyse, especially when we are in front of a very large number of documents. Making correct decisions often requires analysing large volumes of textual information and in spite of this it has more potential to recover crucial information for businesses [11].

In fact, our approach aimed to extract useful knowledge in document collections helping then managers to make better decisions. Regarding data, one of the most important challenges is developing the ability to collect them in real time. Businesses today are increasingly more likely to analyse data coming from the outside: documents, conversations on the web, reports, opinions and many other types of sources. External data are important because, other than providing data in support of governance, they make us never losing touch with reality. These useful data must be integrated with already existing data in the enterprise, past or present: only in this way it is possible to sketch a complete sight of a business system.

Technology. TM and NLP are proposed to extract meaning from data, making visible the hidden meaning. [12, 13]. The attention for opportunities that could arise from the analysis of these type of data, attracts businesses, institutions and society never before in the history of internet. Especially businesses play a central role in this scenario, since they are usually looking for new information to be used as a strategic resource [14]. However, one reason that is having more interest in business education is the ability for companies to monetize this information by selling them to other companies. It is a very interesting phenomenon: companies that produced data analysis software, in the last years innovated and converted their systems offering online and offline solutions; at the same time, we are witnessing to an expansion of market, due to the birth of almost startups.

It is opening a new sector of data analysis dedicated to enterprises that need to be adept in transforming information into knowledge to improve their performances. The flexible and continuous reconnaissance of resources, on the basis of this new knowledge, is the only way to resist the continuous change of the surrounding environment. The term semantic technology is very broad and encompasses a number of techniques used to extract meaning and applications. In this context we refer to the semantic technologies that deal with analysing data in text products from the web, and then we will deal with drawing the boundaries of TM.

Both NLP and TM offer tools to extract non-trivial knowledge from free or unstructured texts typically using Part-Of-Speech (noun, verb, adjective, and so on) and grammatical structures [15].

Actually, a large part of business information appears in unstructured form: emails, letters, reports, transactional documents and financial documents. Business documents are drawn in a specific format and with specific lexicon based on a different type of business. The lexicon used by each enterprise to express ourselves are very singular, so it is necessary that researchers have a good knowledge of the semantic field before proceeding to POS-tagging, remembering that terminological words are semantically univocal for all specialists in a specific sector [16].

As we can observe in Fig. 1 the process of support through documents analysis, which impacts on decision-making, can only develop if it is based on three elements: enterprise, data and technology. In addition, the process does not stop with the decision-making phase: to have effects on the structure of the enterprise and its performance, analysis results are to be shared with each level of the organization not only with top management. This process triggers a cycle of continuous improvement based on knowledge sharing.

Cooperation between these three elements generate a process, as shown in Fig. 1, that lead managers to decision phase. After decisions have to be communicate and shared with each organizational level of enterprise. In this way, data retrieved travel within the enterprise influencing performances. In fact, the last step process, that we called embedding phase, indicates the integration and transfer of document-driven analysis findings to in order to influence and improve business performances.

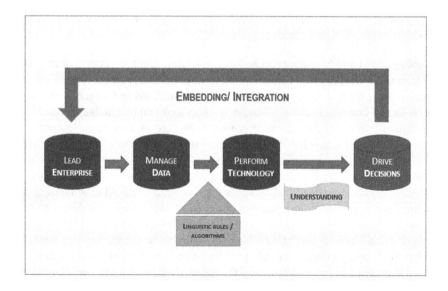

Fig. 1. EDT documents-driven

4 A NooJ Application: Linguistic Resources

NooJ is a linguistic development environment that provides tools to analyse large amount of corpora at several linguistic levels [17–22] For each Linguistic Resource created by linguists, NooJ applied a set of computational formalisms different from most of other NLP software. As mentioned, the goal of our approach is using NooJ to process business documents and to study the use of linguistic phenomena. The main goal is to investigate the semantic aspect showing how to extract the hidden meaning in the texts in an intelligent way: helping manager to make decisions, not replace it.

In fact, the process of decision-making is achieved by combining human assessments with information processed by the system. NooJ provides to the user, through interactive procedures, all the information necessary for the understanding of the problem; it inspects the text from different linguistic level; and even it evaluates the consequences of the linguistic choices made.

A NooJ application to business increases especially the effectiveness of cognitive decisions by managers and this is a revolutionary event, because the goal of IT in the last three decades has been improving efficiency and an intelligent involvement in the resolution of problems. Following previous document-driven DSS approach NooJ could help in the analysis of the company's documentation, not only capturing the underlying meaning in texts but also evaluating the effectiveness of business communication from a linguistic point of view.

Specifically, the business texts are analysed with the aid of the resources within the software environment: in this respect, IT is very important to the development of terminological dictionaries. The terminology was not referred only to the knowledge domain of Economy; it depends on the type of company. Business documents are enriched with technical words that describe well-defined processes and functions: often there are compound words or MWALUs that linguistic analysis based on keywords cannot locate, just because it does not recognize the syntactic, lexical and semantic properties of words. To understand the precise meaning is important to recognize them.

The use of electronic dictionaries is the first approach proposed in order to developing TM analysis based on Linguistic Resources.

To do this we applied Lexicon-Grammar (LG) theoretical and practical framework, which describes the mechanisms of word combinations and provides a complete description of natural language lexical and grammatical structures. LG was based on the works by Maurice Gross during the '60s, and subsequently applied to Italian by Annibale Elia, Maurizio Martinelli and Emilio D'Agostino [23].

Italian electronic dictionaries, developed according to LG, are mainly of two types, considering the formal and semantic aspect of their content. Therefore, we have:

- Simple words dictionaries (DELAS-DELAF) that include all the simple words of Italian, simple word atomic linguistic units and multiword atomic linguistic units;
- Compound words dictionaries (DELAC-DELACF) that include multiword atomic linguistic units and compound words, each string is formed by two or more words and compose a direction unit.

For example, in a previous study [24] with NooJ software we have built an electronic dictionary based on the specific language used in the primordial phases of enterprise setting up, commonly called startup. The following entries defined a supplementary semantic field that enriched the already existing dictionary of Economy.

In Fig. 2, we collected some entries, compound words or MWALUs form. For each entry, firstly we described the category (noun, verb, adjective, and so on), then the internal structure (Part-Of-Speech); the Knowledge domain (predominantly Economy) and finally the inflectional code for Italian Module of NooJ.

By this extract, we observed that every entry of the dictionary are compound words, mostly English. All languages showed a close correlation between terminology and compound words: in fact, terminology needs compound words, especially MWALUs, as it is evidenced by the fact that specialized lexicons include mainly compound words, in some cases up to a number higher than 90% of the listed lexical items. Compound words belonging to specialized lexicons are semantically non-ambiguous, with one unique and validated meaning. Terminological language, by definition, cannot be ambiguous and therefore compound words are the most appropriate linguistic formalization for terminology. It is necessary to reduce maximally the ambiguity of the objects and concepts in the technical-scientific communication.

The four entries highlighted in Fig. 2 provided an example of semantic expansion: the concept of *incubatore* identified a program designed to accelerate the development of startup. If we consider the use of compound words, created placing an adjective (*certificato, privato, pubblico, universitario*) before a name, we defined specifically the meaning of that object, which is also denoted legally and economically in a different way by each entry.

Entry	Category	POS	Domain	FLX
business plan	N	NN	Economy	C779
ciclo di investimento	N	NPN	Economy	C613
coaching management	N	NN	Economy	C779
early stage financing	N	AAN	Economy	C371
early stage investment	N	AAN	Economy	C370
equity crowdfunding	N	NN	Economy	C736
exit strategy	N	NN	Economy	C312
idea imprenditoriale	N	NA	Economy	C544
imprenditore seriale	N	NA	Economy	C343
incubatore certificato	N	NA	Economy	C501
incubatore privato	N	NA	Economy	C501
incubatore pubblico	N	NA	Economy	C501
incubatore universitario	N	NA	Economy	C501
pitch competition	N	NN	Economy	C779
seed financing	N	NN	Economy	C736
startup a vocazione sociale	N	NPNA	Economy	C612
startup innovativa	N	NA	Economy	C569i
venture capital	N	NN	Economy	C779
zombie startup	N	NN	Economy	C779

Fig. 2. An extract of startup electronic dictionary

Recognizing the syntactic, lexical and especially semantic relationship between two or more terms overcomes the keywords research: linguistic analysis results are optimized and informative potential of the document increases. In the following Fig. 3 we provided other examples of semantic expansion, in Agri-food startup communication, represented by a finite state automaton.

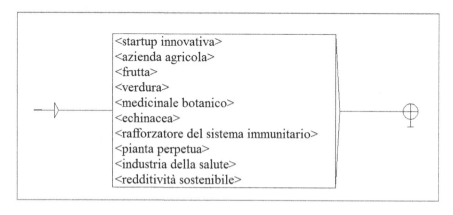

<startup innovativa>
<azienda agricola>
<frutta>
<verdura>
<medicinale botanico>
<echinacea>
<rafforzatore del sistema immunitario>
<pianta perpetua>
<industria della salute>
<redditività sostenibile>

Fig. 3. Finite state automaton in the agri-food language

As syntactic properties, we recognised many compound words in which the second terms of construction (Adjective) defined the specific features of the head (startup innovativa, medicinale botanico, redditività sostenibile).

As far as semantic particularities are concerned, we observed the presence of compound words and MWALUs referring to Biology, Law, AND Chemistry domains. Creating electronic dictionaries is the best way to observe and analyse the use of different terminology in business. The specific lexicon, normally intended for experts, has access to the common language through mass media and very often has a figurative value. The press has always represented the most open channel to spread new words, especially those of English origin. Considering English influence on Italian, it should not therefore be a surprise. Free communication contributed to the internationalization of terms derived from technical-scientific and socio-political areas.

5   Conclusions and Future Works

A company uses documents to communicate, transacts business and analyses its productivity. The rapid growth of data increased the number of documents that almost become unmanageable.

Moreover, business documents made use of a specific terminology, which represents often a limit for analysis of such documents, resulting in the efficiency of DSS that fail to provide immediate answers for managers and knowledge workers.

Processing documents with NLP and TM help companies to overcome this communication gap. Analysing business texts and transforming them into actionable knowledge to spend in decision-making process with help of is one of the best way in which support manager to make decisions.

In order to realise these goals we proposed to introduce in a DSS a natural language interface as NooJ: through Linguistic Resources created, we provided parsers that can be apply to each type of corpus in order to extract the hidden meaning in sentences. The proposed system could be enhanced carrying out texts overcoming ambiguous meaning, with electronic terminological dictionaries in different domains and even create local grammars.

Therefore, a further research could cope with scanning a set of business documents written in natural language and analyse them to understand which kind of information are described. Moreover, we could detected by text, if the concepts expressed are adequate to communicate the business of specific companies. The findings coming from TM analysis and NLP could help companies to make better decisions in terms of product/process, communication and innovation management, knowledge recognition.

References

1. Power, D.J.: Web-based and model-driven decision support systems: concepts and issues. In: AMCIS 2000 Proceedings, p. 387 (2000)
2. Power, D.J.: Decision support systems: a historical overview. In: Burstein, F., Holsapple, C.W. (eds.) Handbook on Decision Support Systems 1, pp. 121–140. Springer, Berlin Heidelberg (2008)
3. Sprague Jr., R.H., Carlson, E.D.: Building Effective Decision Support Systems. Prentice-Hall Inc, Englewood Cliffs (1982)
4. Škrobáčková, M., Kopáčková, H.: Decision support systems or business intelligence: what can help in decision-making? In: Provazniková, R., Kynclová, M. (eds.) Scientific papers of the University of Pardubice, Series D, Faculty of Economics and Administration, Pardubice, Czech Republic, vol. 10, pp. 98–103 (2006)
5. Fedorowicz, J.: Document based decision support in decision support for management. In: Sprague, Jr., R., Watson, H.J. (eds.) Decision Support for Management. Prentice-Hall, Upper Saddle River (1996)
6. Peterson, A.: Document-driven DSS resources. In: DSSResources.COM, December 2000. dssresources.com/dsstypes/docddss.html. Accessed 5 Feb 2016
7. Kopackova, H., Komarkova, J., Sedlak, P.: Decision making with textual and spatial information. WSEAS Trans. Inf. Sci. Appl. 5(3), 259 (2008)
8. Grimes, S.: A Brief History of Text Analytics. B Eye Network. http://www.b-eye-network.com/view/6311. Accessed 24 June 2016
9. Davenport, T.H., Harris, J.G.: Competing on Analytics: The New Science of Winning. Business School Press, Harvard (2007)
10. Gartner Research: Big Data (2013). http://www.gartner.com/it-glossary/big-data/. Accessed 20 March 2016
11. della Volpe, M.: Imprese tra Web 2.0 e Big Data. Nuove frontiere per innovazione e competitività, CEDAM (2013)
12. Radovanovic, M., Ivanovic, M.: Text mining approaches and applications. Novi Sad J. Math. 38(3), 227–234 (2008)
13. Feldman, R., Sanger, J.: The Text Mining Handbook. Cambridge University Press, Cambridge (2007)
14. Blomqvist, E.: The use of semantic web technologies for decision support – A survey. J. Seman. Web 5(3), 177–201 (2014). IOS Press, http://www.semantic-web-journal.net/sites/default/files/swj299_1.pdf
15. Kao, A., Poteet, R.S. (eds.): Natural Language Processing and Text Mining. Springer, London (2007)
16. De Bueriis, G., Elia, A. (eds.): Lessici elettronici e descrizioni lessicali, sintattiche, morfologiche ed ortografiche. Plectica, Salerno (2008)
17. Silberztein, M.: Nooj Manual (2003). http://www.nooj4nlp.net/NooJManual.pdf
18. Silberztein, M.: Corpus linguistics and semantic desambiguation. In: Maiello, G., Pellegrino, R. (eds.) Database, Corpora, Insegnamenti Linguistici, pp. 397–410. Schena Editore/Alain Baudry et C.ie, Fasano/Paris (2012)
19. Silberztein, M.: NooJ Computational Devices. In: Donabédian, A., Khurshudian, V., Silberztein, M. (eds.) Formalising Natural Languages with NooJ, pp. 1–13. Cambridge Scholars Publishing, Newcastle (2013)
20. Silberztein, M.: NooJ V4. In: Koeva, S., Mesfar, S., Silberztein, M. (eds.) Formalising Natural Languages with NooJ 2013, pp. 1–12. Cambridge Scholars Publishing, Newcastle (2013)

21. Silberztein, M.: Analyse et generation transformationnelle avec NooJ. In: Elia, A., Iacobini, C., Voghera, M. (eds.), Livelli di Analisi e Fenomeni di Interfaccia, Rome, Bulzoni (2015)
22. Silberztein, M.: La formalisation des langues: l'approche de NooJ. ISTE Ed, Londres (2015)
23. Elia, A., Martinelli, M., D'Agostino, E.: Lessico e strutture sintattiche: Introduzione alla sintassi del verbo italiano. Liguori Editore, Napoli (1981)
24. Elia, A., Monteleone, M., Esposito, F.: Dictionnaires électroniques et dictionnaires en ligne. Les Cahiers du dictionnaire **6**, 43–62 (2014)

A Decision-Support Tool of Medicinal Plants Using NooJ Platform

Héla Fehri[✉], Mohamed Aly Fall Seideh, and Sondes Dardour

MIRACL Laboratory, Higher Management Institute of Gabes,
University of Gabes, Gabès, Tunisia
hela.fehri@yahoo.fr, almedyfall@gmail.com,
dardour.sondes@yahoo.com

Abstract. In this paper, we are going to develop a tool allowing the identification of recommended medicinal plant names. This tool is based on local grammars built in the linguistic platform NooJ. The developed tool is based on two processes: selection of criteria and identification of the appropriate plant. The first process consists in introducing criteria, such as age, symptoms, sex, contraindication. All these information are inserted in a format developed with Java. The content of this format is saved inside texts which are used as input for the linguistic platform NooJ. The second process consists in parsing the text generated with the procedure of first process, and in identifying the appropriate medicinal plant. This step is based on transducers that represent different rules allowing the recognition of the recommended plant. The system developed is easy to use and does not require computer skills. It gives also consistent results.

Keywords: Medicinal plant · Transducer · Symptom · Contraindication · Disease

1 Introduction

To cure oneself with herbs is a lifestyle. It is important to learn about the allies and proper limitations related to health management. In this context, we should not wait to be very sick, but rather to choose plants that help to restore imbalances as soon as they are felt. It is not a question of replacing the drug by plants, but to define the support they can give to the body in its healing process.

The aim of this paper is to propose a decision-support tool allowing the identification of the recommended medicinal plant name. To develop such system, it is necessary to overcome some challenges. In fact, the number of diseases and plants is very important. Moreover, the symptoms related to some diseases appear to be similar. In this case, it will be difficult to decide about the appropriate plant. For this, it is relevant to define precisely the different criteria and rules to avoid the confusion between diseases symptoms.

Transducers (eventually local grammars) can resolve many problems related to the identification of the appropriate medicinal plant. However, the connection between these transducers and symptoms introduced by a patient is not a trivial task. Therefore, a linguistic platform like NooJ [1, 8] can help us to achieve this task.

© Springer International Publishing AG 2016
L. Barone et al. (Eds.): NooJ 2016, CCIS 667, pp. 246–257, 2016.
DOI: 10.1007/978-3-319-55002-2_21

The remainder of this paper is organized as follows: firstly, a brief overview of the medicinal plants. Secondly, with a description of our proposed method to identify the recommended plant name. Thirdly, with giving a general idea about our resources construction and their implementation in the linguistic platform NooJ. Fourthly, with doing an experimentation and evaluation of the developed tool. Finally, concluding with some perspectives.

2 Herbalism

Herbalism was an honorable profession that laid the foundations of modern medicine, botany, pharmacy, perfumery, and chemistry [2]. Medical herbalism, or simply herbalism or herbology or phytotherapy, is defined by [3] as "the study of herbs and their medicinal uses".

Using herbal medicines as opposed to pharmaceutical products reduce risk of side effects. Herbs typically have fewer side effects and may be safer to use over time. Another advantage to herbalism is the cost. Herbs cost much less than pharmaceutical products.

In herbalism, there are several herbal preparation methods depending on the utilized plant. A medical herbalism can be used as a decoction, infusion, maceration, essential oil or poultices. Also every herb has characteristics such as name [4], family, active ingredient, harvesting, dosage, Preparation (Fig. 1). A medical herbalism can be also classified according to disease [5].

Nom: *Artemisia absinthium*
Famille: Asteracées
Hauteur: 40 à 150 cm
Exposition: Ensoleillée
Sol: Riche et léger
Récolte: Mai à octobre
Formes et préparations: infusions, décoctions

Fig. 1. Information about "mallow"

Health Organization (WHO) estimated that 80% of the world population use herbal medicines as part of their primary health care [6]. Herbalism terminologies are a necessary resource for free users of medicinal plants and phytotherapists. This renewed interest in the natural treatment make the medicinal plants NE recognition as an interesting field of study. In herbalism, plants and diseases names are the most relevant NEs.

As far as we know, there is one work on NEs in herbalism. [7] proposed an identification system of medicinal plants names from French-Arabic parallel corpora.

On the other hand, we have not encountered works on decision-support tool in herbalism using linguistic approaches. However, several organisms, newspapers and associations exist, which include Journal of Applied Research on Medicinal and Aromatic Plants[1], Journal of Medicinal Plant Research[2], International Journal of Phytotherapy[3].

3   Proposed Method

The method that we propose to identify medicinal plants is based on rules. These rules are built manually to express the structure of the information to recognize and take the form of transducers, which will implemented in the linguistic platform NooJ.

The different steps of the proposed method are illustrated in Fig. 2.

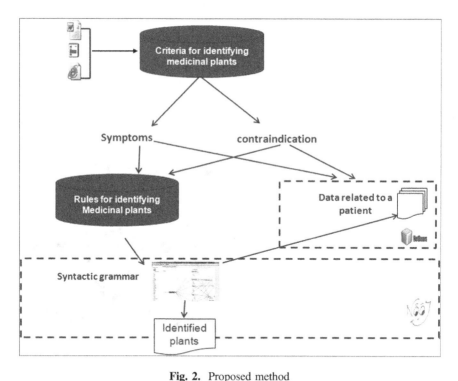

Fig. 2. Proposed method

[1] http://www.sciencedirect.com/science/journal/22147861.

[2] http://www.academicjournals.org/journal/JMPR.

[3] http://www.phytotherapyjournal.com/.

As shown in Fig. 2, the process of identifying medicinal plants is based on two steps: (1) collecting criteria for identifying syntactic patterns and (2) transformation of these patterns to transducers.

In what follows, we will detail these two steps.

3.1   Collection of Criteria for Identifying Syntactic Patterns

This step consists in identifying for each plant a set of criteria. These criteria will be presented below.

Symptoms. The symptom is a sign made by the body in response to a disease or condition. Diseases are typically responsible for many different symptoms, which can be identical from one disease to another.

Examples: diarrhea, tiredness, fever, cold, and so on.

The contraindications

Age. The active ingredient of plants can be toxic to children. In most cases, the EMA (European Medicines Agency) contraindicates these plants for children under twelve and often for those under eighteen. These precautions are especially important for plants administered orally.

Some rare medicinal plants can be used in children under twelve. For example, from the age of six, they can use psyllium seeds (against constipation), ginger (against nausea in children too sick to eat enough) or thyme (against the cough), but always under medical supervision.

Sex. Some plants are totally contraindicated during pregnancy. Others are toxic to the pregnant woman or the fetus.

Chronic diseases and allergy. People who suffer from a chronic illness or allergy should use medicinal plants carefully. In fact, herbal remedies should not aggravate their illness or cause a relapse or side effects or complications, or interact with medications that have been prescribed.

3.2   Identification and Construction of Transducers

In this step, we describe the syntactic patterns using symptoms and contraindications. Then we give the corresponding transducers.

Identification of syntactic patterns. In this step, we assess the understanding and knowledge of drug interactions, contraindications of some medicinal plants in particular for pregnant women, breastfeeding, young children or the elderly and the use of each plant to ensure the safety practice and patient safety.

Based on the properties that characterize each medicinal plant, we focus on the different cases where we can use this plant.

We identified 26 syntactic patterns that allow the identification of the suitable plant. Among these patterns, we cite the following:

<Pattern 1> := **If** (Age = «More than 6 years») ∧ (Sex = Male ∨ Female) ∧ (Symptoms = occasional constipation ∨ Psoriasis) ∧ (¬Renal failure ∧ ¬Liver disease) **Then** Plant = **Aloe**

The pattern 1 describes the conditions that lead to the choice of the plant "aloe". In fact, this plant can help a patient (male or female) whose age is greater than or equal to 6 years and who suffers from occasional constipation or psoriasis on condition he did not kidney failure or disease liver.

<Pattern 2> := **If** (Age = «More than 12 years») ∧ (Sex = Male ∨ Female) ∧ (Symptoms = bronchitis ∨ rheum ∨ cough) **Then** Plant = Cinnamon ∨ Rosemary

The pattern 2 describes the conditions that lead to the choice of plants "Cinnamon" and "Rosemary". Indeed it can help a patient whose age is above 12 years suffering from bronchitis or colds or cough on condition he did have ulcers.

<Pattern 3> := **If** (Sex = Male ∨ Female) ∧ (Symptom = Nosebleed) ∧ (1 pregnancy ∧ 1 breastfeeding ∧ 1 heart or kidney disease) **Then** Plant = Horsetail

The pattern 3 describes the conditions that lead to the choice of the plant "Horsetail". In fact, these plants can help a patient (male or female) of any age and suffering from nosebleed provided that HE does not have a heart or kidney disease and if it is a woman, she should not be pregnant or breastfeeding.

Transformation of syntactic patterns into transducers. This step consist in formalizing the rules already built in the previous step using transducers. Figure 3 represents the main transducer for identifying the corresponding medicinal plant.

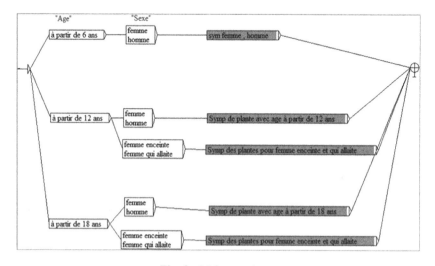

Fig. 3. Main transducer

This type of representation has at least three advantages:

- Recursion calls between graphs and sub-graphs.
- The possibility of adding conditions on transitions.
- The ability to associate outputs to the transitions.

There are four graphs that represent the different types of medicinal plants "sym femme, homme", "Symp de plante avec age à partir de 12 ans", "Symp des plantes pour femme enceinte et qui allaite", "Symp de plante avec age à partir de 18 ans".

Let us note that the transducer of Fig. 3 contains 34 sub-graphs. These sub-graphs show a detailed description for the different criteria involved in the identification of suitable plant. To illustrate these criteria, we detail in the following sub-graphs that are embedded in the node "sym de plante avec age à patir de 18 ans."

The transducer of Fig. 4 describes the different criteria involved in the identification of suitable plant. Each plant contains contraindications such as allergy or chronic disease. To illustrate this, we detail in Fig. 5 the contraindications related to the plant «Juniper» (Genévrier).

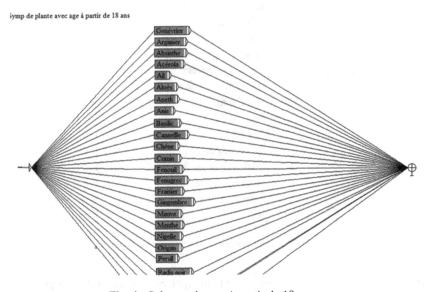

Fig. 4. Sub-transducer «A partir de 18 ans»

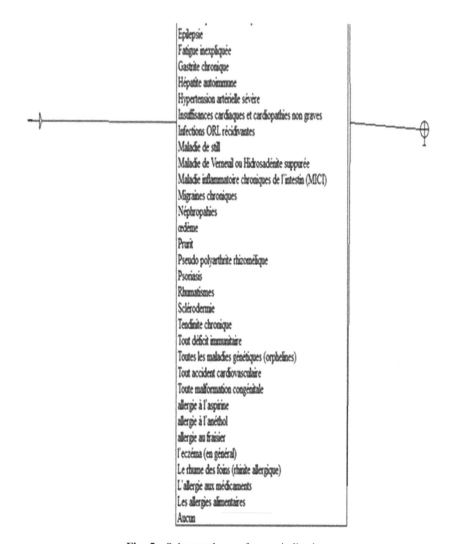

Fig. 5. Sub transducer of contraindication

If the age is «à partir de 18 ans» (more than eighteen years) and the symptom is one of the list defined into sub-transducer in Fig. 6, the appropriate plant will be «Genévrier» (Junipe).

4   Experimentation and Evaluation

We achieved the experimentation of our linguistic resources with NooJ. As mentioned above, NooJ uses already built syntactic grammars. In this section, we present some interfaces to explain how to use our developed tool. The main interface of our tool is illustrated in Fig. 7.

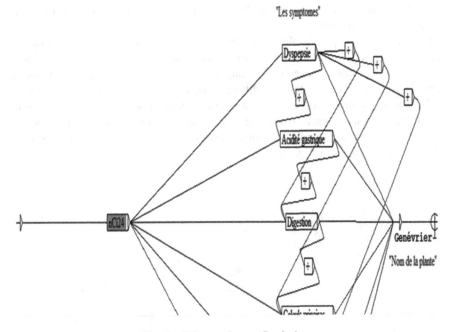

Fig. 6. Sub transducer «Genévrier»

Fig. 7. Main interface

As shown in Fig. 7, the main interface allows accessing to two different profiles: Administrator and patient.

The administrator profile allows the management of symptoms and contraindications. Also, the administrator must be authenticated using a user name and password.

The patient profile is accessible to any user trying to use the help of medicinal plants. In what follows, we describe different steps a patient should be following.

The first step consists in the selection of criteria that allows the identification of the appropriate plant name such as the age, the symptoms, the sex, the contraindication (e.g. if it is a pregnant woman). All these criteria are introduced in a form developed with Java as illustrated in Fig. 8.

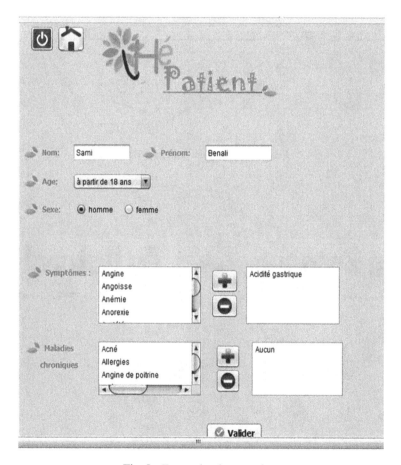

Fig. 8. Form related to a patient

The patient must enter all required fields on the form of the Fig. 8.. All the data will be saved in a file that will be the input of the linguistic platform NooJ.

The second step consists in parsing the text generated with the first step and identifing the appropriate medicinal plant. This step is based on transducers that represent different rules allowing the recognition of the recommended plant. In fact, the transducers described in Figs. 3, 4, 5 and 6 will be applied on this file to display the appropriate plant as illustrated in Fig. 9.

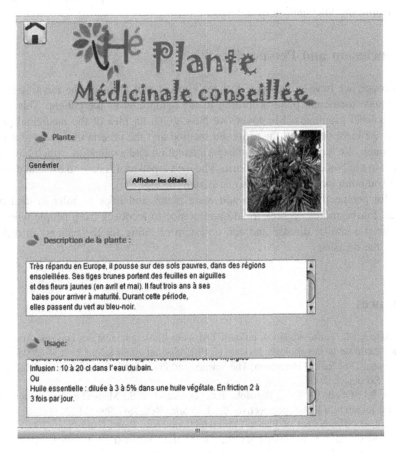

Fig. 9. Displayed result

As shown in Fig. 9, the identified plant is displayed in the Java interface with some informations. These informations concern:

- The image of the plant.
- A plant description (scientific name, family, stem, flowers, height, where it can be found, … and so on).
- Usage: describe the mode of plant preparation (decoction, infusion, maceration…).

Let us note that the connection between Java and NooJ is achieved with AutoHotkey[4].

Our method provides 80% of well-identified plant name. The obtained result is promising but also shows that there are some non-resolved problems. In fact, some cases require the intervention of a doctor.

5   Conclusion and Perspectives

In this paper, we have proposed a decision-support tool allowing the identification of the adequate medicinal plant according to the symptoms of the patient. This tool is based on local grammars. Moreover, we have given an idea of the medicinal plants. Besides, we have described the proposed method and the criteria that should be taken into account. We have also given an experimentation and evaluation, which have been achieved in the NooJ linguistic platform. The developed tool is easy to use and does not require computer skills. It gives also consistent results.

As for perspectives, we aim to add other plants and rules in order to treat other diseases. Furthermore, we seek to add a reference to feedback from people who have experienced a similar disease and add a system of rating for the patients in order to improve the decision.

References

1. Silberztein, M., Tutin, A.: NooJ, un outil TAL pour l'enseignement des langues. Application pour l'étude de la morphologie lexicale en FLE. Spécial Atala 8(2), 123–134 (2005)
2. Hoffmann, D.: Medical Herbalism: The Science and Practice of Herbal Medicine. Haling Arts Press, Rochester (2003)
3. Ameh, S.J., Obodozie, O.O., Babalola, P.C., Gamaniel, K.S.: Medical herbalism and herbal clinical research: a global perspective. Br. J. Pharm. Res. 1(4), 99–123 (2011)
4. Gledhill, D.: The Names of Plants. Cambridge University Press, Cambridge (2008)
5. Fabre, M-C., Genin, A., Merigoux, J., Moget, E.: Herboristerie Familiale. Des recettes simples avec des plantes simples pour résoudre les problèmes simples (1992)
6. Lim-Cheng, N.-R., Richmond, C., Co, J., Gaudiel, C.H.S., Umadac, D.F., Victor, N.L.: Semi-automatic population of ontology of Philippine medicinal plants from on-line text. Presented at the DLSU Research Congress, De La Salle University, Manila, Philippines, 6–8 March 2014

---

[4] https://autohotkey.com/docs/Tutorial.htm.

7. Seideh, M.A.F., Fehri, H., Haddar, K.: Named entity recognition from Arabic-French herbalism parallel corpora. In: Okrut, T., Hetsevich, Y., Silberztein, M., Stanislavenka, H. (eds.) NooJ 2015. CCIS, vol. 607, pp. 191–201. Springer, Heidelberg (2016). doi:10.1007/978-3-319-42471-2_17
8. Silberztein, M.: Formalizing Natural Languages: The NooJ Approach. Wiley, London (2016)

Author Index

Printed in the United States
By Bookmasters